T0297004

CAMBRIDGE LIBRARY COLLECTION

Books of enduring scholarly value

Mathematical Sciences

From its pre-historic roots in simple counting to the algorithms powering modern desktop computers, from the genius of Archimedes to the genius of Einstein, advances in mathematical understanding and numerical techniques have been directly responsible for creating the modern world as we know it. This series will provide a library of the most influential publications and writers on mathematics in its broadest sense. As such, it will show not only the deep roots from which modern science and technology have grown, but also the astonishing breadth of application of mathematical techniques in the humanities and social sciences, and in everyday life.

The New Quantum Mechanics

George Birtwistle (1877–1929) published The New Quantum Mechanics in 1928. His stated aim was to give a detailed account of work which had brought the relatively new subject of quantum mechanics to the fore in the previous few years. The earlier chapters give a restatement of Alfred Landé's theory of multiplets which reconciles it with the new mechanics which follow. Later chapters present the matrix theory of Heisenberg, the q-number theory of Dirac and the wave mechanics of Schroedinger, and sythesise new theories, statistics and controversies in the work of de Broglie, Bose, Einstein, Fermi and Dirac. The book gives a complete overview of the state of quantum mechanics at the end of the second decade of the twentieth century, making it a valuable benchmark for historians of science and mathematicians alike.

Cambridge University Press has long been a pioneer in the reissuing of out-of-print titles from its own backlist, producing digital reprints of books that are still sought after by scholars and students but could not be reprinted economically using traditional technology. The Cambridge Library Collection extends this activity to a wider range of books which are still of importance to researchers and professionals, either for the source material they contain, or as landmarks in the history of their academic discipline.

Drawing from the world-renowned collections in the Cambridge University Library, and guided by the advice of experts in each subject area, Cambridge University Press is using state-of-the-art scanning machines in its own Printing House to capture the content of each book selected for inclusion. The files are processed to give a consistently clear, crisp image, and the books finished to the high quality standard for which the Press is recognised around the world. The latest print-on-demand technology ensures that the books will remain available indefinitely, and that orders for single or multiple copies can quickly be supplied.

The Cambridge Library Collection will bring back to life books of enduring scholarly value across a wide range of disciplines in the humanities and social sciences and in science and technology.

The New Quantum Mechanics

George Birtwistle

CAMBRIDGE UNIVERSITY PRESS

Cambridge New York Melbourne Madrid Cape Town Singapore São Paolo Delhi

Published in the United States of America by Cambridge University Press, New York

www.cambridge.org
Information on this title: www.cambridge.org/9781108005326

This edition first published 1928
This digitally printed version 2009

ISBN 978-1-108-00532-6

THE NEW QUANTUM MECHANICS

Cambridge University Press
Fetter Lane, London

New York
Bombay, Calcutta, Madras
Toronto
Macmillan

Tokyo
Maruzen-Kabushiki-Kaisha

THE NEW QUANTUM MECHANICS

by

GEORGE BIRTWISTLE,
FELLOW OF PEMBROKE COLLEGE,
CAMBRIDGE

CAMBRIDGE
AT THE UNIVERSITY PRESS
1928

PRINTED IN GREAT BRITAIN

PREFACE

This book is concerned with the development of quantum mechanics during the past two years. A detailed account is given of the matrix theory of Heisenberg, Born and Jordan, the q-number theory of Dirac, and the wave mechanics of Schrödinger. The earlier chapters are devoted to a restatement of the Landé theory of the multiplets in a form which is in consonance with the new mechanics which is to follow; some later chapters are given up to the de Broglie theory of particles and their waves, and to the new statistics of Bose, Einstein, Fermi and Dirac.

The book closes with the resonance theory of the helium spectrum lately given by Heisenberg, and with the new speculations of Bohr on the limitations imposed by the quantum theory upon the possibilities of experimental observation.

G. B.

HOTEL PHOENIX,
COPENHAGEN

1 *October* 1927

CONTENTS

THE NEW QUANTUM MECHANICS

CHAPTER I

THE ORIGIN AND DEVELOPMENT OF THE NEW QUANTUM MECHANICS

1. *The origin of the new quantum mechanics* was an epoch-making memoir by Werner Heisenberg[1] which contained the new concept which was to lead to the phenomenal developments of quantum mechanics of the past two years. Up to this time the quantum theory (the 'older' quantum theory) postulated the existence of stationary states of the atom calculated by the use of the classical mechanics and selected by the use of quantum conditions satisfied by the action variables of that theory. In the new mechanics the equations have the same form as in the classical theory, but the variables no longer satisfy the commutative law of multiplication, that is, xy is not in general equal to yx; the quantum conditions of the older theory are replaced by equations which enable the difference $xy - yx$ to be calculated; these equations involve Planck's constant h.

For some years before 1925, Sommerfeld, Heisenberg, Landé and Pauli[2] had been grappling with the complex problem of the multiplets and their Zeeman separations. By the use of a system of quantum numbers l, s, j connected with the respective angular momenta of the series electron, the core, and the whole atom, they had given a qualitative account of the multiplets of the alkalis,

[1] W. Heisenberg, Zs. f. Phys. **33**, p. 879, July **1925**.

[2] Various papers in the Zs. f. Phys. from **1922** to **1924**.

alkaline earths, etc., the work culminating in a very general empirical formula—the g-formula of Landé—which enabled the Zeeman separations of a multiplet to be worked out quantitatively in terms of the quantum numbers l, s, j. An essential part of the scheme was the use of *half odd integers* as well as integers as possible values of l, s, j and this formed no part of the older quantum theory. Heisenberg's new theory however at once led to the formula $\left(n + \frac{1}{2}\right) h\nu$ as the energy of a stationary state of Planck's oscillator, so that half odd integers came quite naturally into the new results.

A real difficulty too had been met with in the spectrum of neutral helium, where *two* electrons revolve round the nucleus (the simplest many electron problem), all the theoretical results found being at variance with experiment; again in the problem of the 'crossed' fields, where an atom is exposed to the combined action of electric and magnetic fields, fundamental difficulties arose.

But the work which directly led to the formulation of the new mechanics was that of Kramers and Heisenberg[1] on dispersion.

They worked out the absorption and scattering of radiation by an atom regarded as a multiply periodic system of the classical theory, perturbed by the incident radiation. They thus found a result in terms of the orbital frequencies and amplitudes. This result was then translated, by the use of the correspondence principle, into one containing the experimentally observable magnitudes, namely, the frequencies and amplitudes of the spectral lines emitted by the atom[2].

Heisenberg then sought to develop a scheme of quantum kinematics by which the quantum formulae would be ob-

[1] H. A. Kramers and W. Heisenberg, Zs. f. Phys. **31**, p. 681, **1925**.
[2] 'The Quantum Theory of the Atom,' by G. Birtwistle, Cambridge, **1926**, §§ 148 to 154; this book will be referred to as Q.T.A.

tained directly in terms of these experimentally observable magnitudes, the frequencies and intensities of the spectrum, without the intermediate use of orbital frequencies and amplitudes, which by their nature can never probably be observed.

This meant that instead of representing a dynamical quantity, as on the classical theory, by a one-dimensional *line* of terms (a Fourier series) of the type

$$C_1 e^{2\pi i(\omega t)},\ C_2 e^{2\pi i(2\omega t)},\ \ldots\ C_n e^{2\pi i(n\omega t)},\ \ldots$$

it should, on the quantum theory, be represented by a two-dimensional *table* of terms (a matrix) of the type

$$\begin{pmatrix} A\,(1,1)\,e^{2\pi i\nu(1,1)t}, & A\,(1,2)\,e^{2\pi i\nu(1,2)t}, & \ldots \\ A\,(2,1)\,e^{2\pi i\nu(2,1)t}, & A\,(2,2)\,e^{2\pi i\nu(2,2)t}, & \ldots \\ \ldots & \ldots & \ldots \end{pmatrix}$$

where $A\,(n, m)\,e^{2\pi i\nu(n,m)t}$ represents the spectral line of frequency $\nu\,(n, m)$ and intensity dependent upon $A\,(n, m)$ due to a Bohr transition from the stationary state n to the stationary state m.

Heisenberg, Born and Jordan then proceeded to develop the matrix mechanics, and in two papers[1] worked out the theory of the harmonic and anharmonic oscillators, gave a perturbation theory for non-degenerate and degenerate systems, and with it a direct deduction of the dispersion formula and of formulae required for the calculation of Zeeman intensities.

While this matrix theory was being developed, Dirac[2] was working out a theory on somewhat different lines; he discovered that the quantum conditions for a multiply periodic system could be expressed in terms of the 'Poisson brackets' of the classical mechanics. As in the classical mechanics the only differential coefficients essential to the theory can be put into Poisson bracket form, the diffi-

[1] M. Born and P. Jordan, Zs. f. Phys. **34**, p. 858, Sept. **1925**. M. Born, W. Heisenberg and P. Jordan, Zs. f. Phys. **35**, p. 557, Nov. **1925**. These papers will be referred to as Q.M. I and Q.M. II respectively.

[2] P. A. M. Dirac, Proc. Roy. Soc. A. **109**, p. 642, Nov. **1925**.

culties met with by Born, Heisenberg and Jordan in their efforts to preserve the Hamiltonian form of the equations of motion by a suitable form of matrix differentiation, are not encountered on Dirac's theory. Dirac goes further and shows that we can work in the new mechanics without using matrices; he calls functions of the dynamical variables q-numbers (which do not obey the commutative law of multiplication) and ordinary numbers (which do) c-numbers; though in the interpretation of q-number results in terms of experiment (where c-numbers must be used) the matrix of the q-number is necessary.

In this way Dirac[1] gave the theory of the hydrogen atom on the new mechanics, a crucial test of the new system, as the Bohr formula is the basis of the whole structure of quantum mechanics; he found that the term form $\frac{Rhc}{n^2}$, with n equal to an integer 1, 2, ... is preserved in the new mechanics, as had been simultaneously shown by Pauli[2] who used the Heisenberg matrices.

For some time before this, serious difficulties had existed in the interpretation of the multiplets, their Zeeman separations and the corresponding X-ray multiplets. The function assigned to the core in the Landé (l, s, j) model, Pauli[3] urged should be assigned to the series electron, so that every electron has four independent quantum numbers. He further laid down the axiom that no two electrons in the atom can have all four quantum numbers the same; this axiom, known as the 'exclusion' principle of Pauli[4] (or the Pauli 'verbot'), at once accounted for the maximum number of electrons 2, 8, 18, 32, ... in the K, L, M, N, ... 'shells' in accord with Bohr's theory of the atomic structure of the elements.

[1] P. A. M. Dirac, Proc. Roy. Soc. A. **110**, p. 561, Jan. **1926**.
[2] W. Pauli, Zs. f. Phys. **36**, p. 336, Jan. **1926**.
[3] W. Pauli, Zs. f. Phys. **31**, p. 373, Dec. **1924**.
[4] W. Pauli, Zs. f. Phys. **31**, p. 765, Jan. **1925**.

In October 1925, Uhlenbeck and Goudsmit[1] put forward their theory of the spinning electron which enabled the quantum number s previously associated with the angular momentum of the core electrons to be associated with the spin of the series electron, thus giving a mechanical interpretation to the transfer of the quantum number s from the core to the electron, of which Pauli had postulated the theoretical necessity. In March 1926, Heisenberg and Jordan[2], using the new mechanics and the spinning electron theory of Uhlenbeck and Goudsmit, calculated the fine structure of the doublets of the alkalis, their Zeeman separations and intensities, and completely cleared up the difficulties with which this problem had been beset for many years. They proved the Landé g-formula and also the Sommerfeld formula for the Paschen-Back changes.

Early in 1926 the complexity of the matrix procedure had made itself felt, with its infinite number of difference equations with an infinite number of unknowns, and several writers endeavoured to bring the new mechanics within the range of more highly developed analysis. Lanczos[3], with his 'field theory,' had brought it into contact with the theory of integral equations; Born and Wiener[4] had devised an 'operator calculus,' but just about this time a most remarkable development of the theory on lines totally different from those of Heisenberg and Dirac was put forward by Schrödinger[5]. Fired by the new ideas of Louis de Broglie[6] on material particles and their associated waves, he assumed that the dynamics of an atom cannot be represented by a point moving through

[1] G. E. Uhlenbeck and S. Goudsmit, Naturwissensch. **47**, p. 953, Nov. **1925**; Nature, **117**, p. 264, Feb. **1926**.
[2] W. Heisenberg and P. Jordan, Zs. f. Phys. **37**, p. 263, March **1926**.
[3] K. Lanczos, Zs. f. Phys. **35**, p. 812, Feb. **1926**.
[4] M. Born and N. Wiener, Zs. f. Phys. **36**, p. 174, Jan. **1926**.
[5] E. Schrödinger, Ann. der Phys. **79**, p. 361, Jan. **1926**; **79**, p. 489, Feb. **1926**; **79**, p. 734, March **1926**.
[6] L. de Broglie, Ann. de Phys. **10**, p. 22, **1925** (Thèses, Paris **1924**).

the coordinate space (the q-space) as in the classical theory, but must be represented by a wave in that space, and obtained a differential equation which the 'wave function' ψ must satisfy. This equation contains E, the total energy of the atom. In general, this equation only has solutions (which are continuous, unique and bounded in the q-space) for certain definite values of E, viz. $E_1, E_2, \ldots$ ('eigenwerte'), and the corresponding solutions $\psi_1, \psi_2, \ldots$ are the 'eigenfunctions.' The eigenwerte are the energy levels of the atom, and Schrödinger shows how the eigenfunctions may be used to determine the Heisenberg matrices by a process of quadratures. The strength of Schrödinger's 'wave mechanics' lies in the fact that it brings the new mechanics within the scope of the highly developed analysis of differential equation theory, and then reduces the calculation of the intensities of the lines to a series of integrations.

In March and April 1926 appeared two further papers by Dirac. In the former[1] he obtained the angular momentum relations of Heisenberg, Born, and Jordan[2] by his Poisson bracket methods, and using his angle and action variable theory developed for q-numbers, found the g-formula of Landé, and also Kronig's results for the relative intensities of a multiplet and their components in a weak magnetic field.

In the latter[3] he extended his theory to relativity mechanics and used it to give the theory of the Compton effect; he obtained a more consistent agreement with Compton's experimental results than did Compton by his theory of light quanta.

In May and June 1926 two further memoirs by Schrödinger[4] appeared. In the former he develops a theory of

[1] P. A. M. Dirac, Proc. Roy. Soc. A. **111**, p. 281, March **1926**.
[2] Q.M. II.
[3] P. A. M. Dirac, Proc. Roy. Soc. A. **111**, p. 405, April **1926**.
[4] E. Schrödinger, Ann. der Phys. **80**, p. 437, May 1926; **81**, p. 109, June **1926**.

perturbations and applies it to find the *intensities* of the lines in the Stark effect and so calculates the 'total intensities' of the lines H_α, H_β, H_γ, H_δ; in the latter he develops a dispersion theory and obtains the Kramers-Heisenberg formula.

By this time the computational value of the Schrödinger methods had been appreciated by earlier workers. In June and July 1926, Born[1] published two papers on collision phenomena; he finds a solution of the Schrödinger wave equation consisting of incident plane waves representing the approaching electron, and these waves are scattered by the atomic system. He assumes that the square of the amplitude of the wave scattered in any direction determines the probability of the electron being scattered in that direction, with an energy given by the frequency of the wave.

In June 1926 Heisenberg[2] wrote an outstanding paper on *resonance* in atoms with two electrons, which contained the key to the solution of the spectrum of neutral helium, with its mystery of the ortho and para helium terms. It is well known that the latter is a single term system containing the 'ground' term, and the former a doublet system; also that the terms of the one system do not combine with those of the other.

All these characteristics were accounted for, the difference between the ortho and para terms corresponding to the frequency of the energy pulsations from the one electron to the other within the atom. This work is carried out by the use of his matrix mechanics, but for the first time he uses Schrödinger's calculus to confirm his theory of the non-combination of the ortho and para helium terms.

In a second paper in July[3] he computed the ortho and para helium separations by the use of Schrödinger's calculus and found results of the right order of magnitude both for

[1] M. Born, Zs. f. Phys. **37**, p. 863, June **1926**; **38**, p. 803, July **1926**.
[2] W. Heisenberg, Zs. f. Phys. **38**, p. 411, June **1926**.
[3] W. Heisenberg, Zs. f. Phys. **39**, p. 499, July **1926**.

He and Li+, though it is evident that a more refined perturbation theory is required in quantum mechanics with elaborations on the lines used in astronomy.

In August 1926, Dirac[1] gave an illuminating account of the derivation of the Heisenberg matrices by the use of Schrödinger's eigenfunctions and considered the two electron problem for the atom by the Schrödinger method; he found that there were two solutions satisfying the equations, and that one led to the Pauli verbot and the other to the Bose-Einstein statistical mechanics. He also worked out a perturbation theory for the wave mechanics of Schrödinger in which the approximation is carried to the second order, and expressions for the Einstein *B* coefficients are found.

In December 1926, Dirac[2] obtained a general formal matrix theory using operational methods, and derived Schrödinger's equation as part of this matrix calculus; he also generalised it for cases where the Hamiltonian contains the time explicitly.

In the first of two later papers (1927) Dirac[3] gave a new theory of emission and absorption of radiation. He considers the interaction of an assembly of light quanta with an atom and finds expressions for both of the Einstein *A* and *B* coefficients; in the second he gives a theory of dispersion and deduces the Kramers-Heisenberg formula.

It is hardly possible to pass on without remarking upon the almost uncanny anticipation by Courant and Hilbert[4] of the pure mathematical theory required for the new mechanics, which has had so much to do with the rapidity of its development.

[1] P. A. M. Dirac, Proc. Roy. Soc. A. **112**, p. 661, August **1926**.
[2] P. A. M. Dirac, Proc. Roy. Soc. A. **113**, p. 621, Dec. **1926**.
[3] P. A. M. Dirac, Proc. Roy. Soc. A. **114**, p. 243, Feb. **1927**; A. **114**, p. 710, April **1927**.
[4] 'Methoden der mathematischen Physik,' **1**, by R. Courant and D. Hilbert, Berlin, **1924**.

CHAPTER II

THE MULTIPLETS OF SERIES SPECTRA AND THE l-s-j SCHEME OF LANDÉ

2. *Series spectra and their multiplets.*

The 'terms' of a series[1] are of the form $\frac{Rhc}{n^2} f(n, l)$, where $f(n, l) \to 1$ as $n \to \infty$, n being the principal quantum number and l a second quantum number associated with the series to which the term belongs. Writing the above expression as $\phi(n, l)$, the terms can be written as

$\phi(n, 0)$, $\phi(n+1, 0)$, $\phi(n+2, 0)$, ..., where $l = 0$ (s terms)

$\phi(n+1, 1)$, $\phi(n+2, 1)$, ..., where $l = 1$ (p terms)

$\phi(n+2, 2)$, ..., where $l = 2$ (d terms)

$\phi(n+3, 3)$, ..., where $l = 3$ (b terms)

....

[For sodium, for example, where in the normal state the series electron is in the M shell, $n = 3$.]

The above terms are usually written as

$$\left.\begin{array}{llll} 1s, & 2s, & 3s, & \dots \\ & 2p, & 3p, & \dots \\ & & 3d, & \dots \\ & & & 4b, \dots \end{array}\right\} \text{(\textit{Term} series)}$$

and the various series *lines* are given by the transitions

$Np \to 1s$, $N = 2, 3, 4, \dots$, the *principal* series
$Ns \to 2p$, $N = 2, 3, 4, \dots$, the *sharp* series
$Nd \to 2p$, $N = 3, 4, 5, \dots$, the *diffuse* series
$Nb \to 3d$, $N = 4, 5, 6, \dots$, the *Bergmann* series
(*Line* series)

[1] Q.T.A. chap. XI.

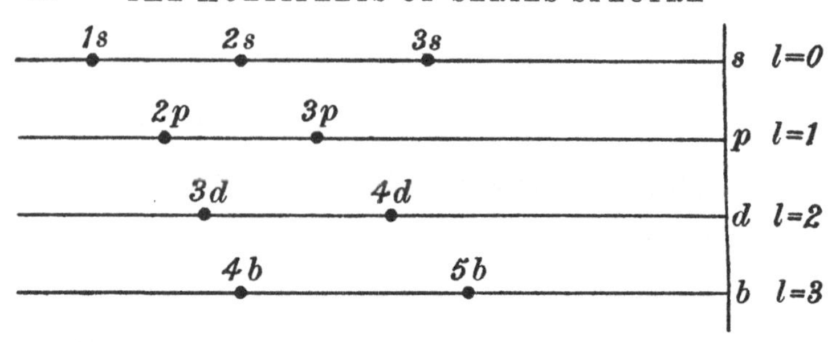

The terms can be represented, after Bohr, by the above diagram where the horizontal distance of a term point from the vertical line measures its energy value.

3. *Multiplets.*

The series lines when examined with high power are found in general to be multiple in structure. Thus in the case of the alkalis, the s terms remain single, but the $p, d, b, \ldots$ terms each split up into two (a doublet) so that the above figure becomes

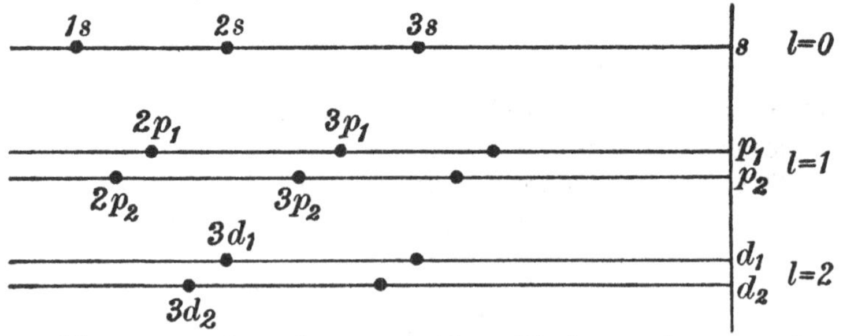

The separations become wide with increasing atomic number, the magnitude for the $p_1 p_2$ levels ranging from 6 Å for sodium to 422 Å for caesium.

Sommerfeld and Landé introduced a third quantum number j so that a pair of corresponding p_1, p_2 terms had $l = 1$, but different values of j.

4. *Earlier theory of the multiplets.*

In the earlier form of the theory the field acting on the

electron was supposed to a first approximation to be central, which leads to a plane periodic motion on which is superposed a rotation in its own plane[1]. The angular momentum (L) due to this rotation was associated with the quantum number l, the angular momentum (S) due to the rest of the atom ('core') with a quantum number s, and the resultant (J) of L and S (the resultant angular momentum of the whole atom) with the quantum number j.

It will be seen later (§ 76) that the new Heisenberg mechanics requires L, S, J to be respectively equal to

$$\frac{h}{2\pi}\sqrt{l(l+1)},\ \frac{h}{2\pi}\sqrt{s(s+1)},\ \frac{h}{2\pi}\sqrt{j(j+1)},$$

where h is Planck's constant.

On account of the coupling forces between the series electron and the core, the triangle SLJ precesses about J.

The correspondence principle then leads to the selection rules[2] that in a transition, l can only change by ± 1, while j may change by ± 1 or 0.

Further since J is the resultant of L and S, and must therefore lie between $L+S$ and $|L-S|$, the equation $|l-s| \leqslant j \leqslant l+s$ gives limits for j.

If $s=0$, then j must $=l$ (singulet terms).

If $s=\frac{1}{2}$, then $j=l-\frac{1}{2}$ or $l+\frac{1}{2}$ (doublets), except for $l=0$.

$$\therefore\quad l=0 \text{ gives } j=\frac{1}{2},$$

$$l=1 \quad ,, \quad j=\frac{1}{2} \text{ or } \frac{3}{2},$$

$$l=2 \quad ,, \quad j=\frac{3}{2} \text{ or } \frac{5}{2}.$$

[1] Q.T.A. chap. VIII. [2] Q.T.A. § 87.

If $s = 1$, then $j = l - 1$, l or $l + 1$ (triplets), except for $l = 0$, so that

$l = 0$ gives $j = 1$, $\quad l = 1$ gives $j = 0$, 1 or 2,
$l = 2$,, $j = 1$, 2 or 3, $\quad l = 3$,, $j = 2$, 3 or 4.

If $s = \frac{3}{2}$, $j = l - \frac{3}{2}$, $l - \frac{1}{2}$, $l + \frac{1}{2}$, $l + \frac{3}{2}$ (quartets), except for $l = 0$ or 1, so that

$l = 0$ gives $j = \frac{3}{2}$, $\quad l = 1$ gives $j = \frac{1}{2}, \frac{3}{2}$ or $\frac{5}{2}$,
$l = 2$,, $j = \frac{1}{2}, \frac{3}{2}, \frac{5}{2}$ or $\frac{7}{2}$, $\quad l = 3$,, $j = \frac{3}{2}, \frac{5}{2}, \frac{7}{2}$ or $\frac{9}{2}$.

These results are embodied in the following tables due to Landé; the symbols are different from those used by Landé, having been altered to bring them into line with the new mechanics to be developed later. The quantum numbers l, s, j used here are respectively $K - \frac{1}{2}$, $R - \frac{1}{2}$, $J - \frac{1}{2}$, where K, R, J are Landé's quantum numbers.

5. *Landé's scheme of quantum numbers for the multiplet terms.*

Singulets

l \ j	0	1	2	3	...
0	s				
1		p			
2			d		
3				b	
...					...

$(s = 0)$

Doublets

l \ j	$\frac{1}{2}$	$\frac{3}{2}$	$\frac{5}{2}$	$\frac{7}{2}$	...
0	s				
1	p_1	p_2			
2		d_1	d_2		
3			b_1	b_2	
...					...

$\left(s = \frac{1}{2}\right)$

Triplets

l \ j	0	1	2	3	4	...
0		s				
1	p_1	p_2	p_3			
2		d_1	d_2	d_3		
3			b_1	b_2	b_3	
...						...

$(s=1)$

Quartets

l \ j	$\frac{1}{2}$	$\frac{3}{2}$	$\frac{5}{2}$	$\frac{7}{2}$	$\frac{9}{2}$	...
0		s				
1	p_1	p_2	p_3			
2	d_1	d_2	d_3	d_4		
3		b_1	b_2	b_3	b_4	
...						...

$\left(s=\frac{3}{2}\right)$

This scheme applies to any multiplet and is independent of the principal quantum number n (Preston's rule).

This empirical scheme obtained as the result of long and laborious research by Landé is justified by the exactness with which it enables the anomalous Zeeman separations of the lines of a multiplet to be calculated.

CHAPTER III

THE NORMAL AND ANOMALOUS ZEEMAN EFFECT; THE LANDÉ *g*-FORMULA

6. *The normal Zeeman effect of the singulets.*

A magnetic field splits each spectral line into three when the line is viewed at right angles to the field; the central one is polarised parallel to the field and the two outer ones at right angles to the field; the central one is at the place of the original line and the outer ones equidistant from it. The effect of the field H is to cause the Larmor precession[1] of the orbit of the electron about the axis of the field. The frequency of this precession is o, where $o = \frac{eH}{4\pi mc}$.

If E' is the increase of energy due to o, then Bohr's formula[2] $dE = \omega . dI$, which is here $dE' = o . dM$, where M is the action variable corresponding to the frequency o, leads to $E' = o . M$. Writing $M = mh$, in the usual manner, where m is a quantum number, $E' = omh$.

If E_0 is the energy in the absence of the magnetic field and E the energy in presence of the field,

$$E = E_0 + E' = E_0 + moh.$$

The frequency ν due to a transition where m changes to m' is given by

$$\nu = \frac{1}{h}[(E_0 + omh) - (E_0' + omh)]$$

$$= \frac{E_0 - E_0'}{h} + o(m - m').$$

If ν_0 is the frequency of the original line,

$$\nu_0 = \frac{E_0 - E_0'}{h},$$

so that

$$\nu - \nu_0 = o(m - m').$$

[1] Q.T.A. § 77. [2] Q.T.A. § 31.

The possible transitions are given by the correspondence principle[1], which shows that m changes by $\pm$ 1 or 0, i.e. $|\Delta m| = 1$ or 0, and that if $|\Delta m| = 1$ there is circular polarisation in a plane normal to the field, seen as linear polarisation at right angles to the field when viewed across it. Also if $|\Delta m| = 0$, there is linear polarisation parallel to the field.

Hence $\nu - \nu_0 = \pm\, o$ or 0, the normal Zeeman triplet; the magnitude of the separation of the outer lines from the central one is $\Delta\nu = o$.

7. *Slow precession of the angular momentum vector about an axis.*

If Ox, Oy, Oz are fixed, the energy E_0 of a system of particles is

$$E_0 = \tfrac{1}{2}\Sigma\mu\,(\dot{r}^2 + r^2\dot{\theta}^2 + r^2\sin^2\theta\,\dot{\phi}^2).$$

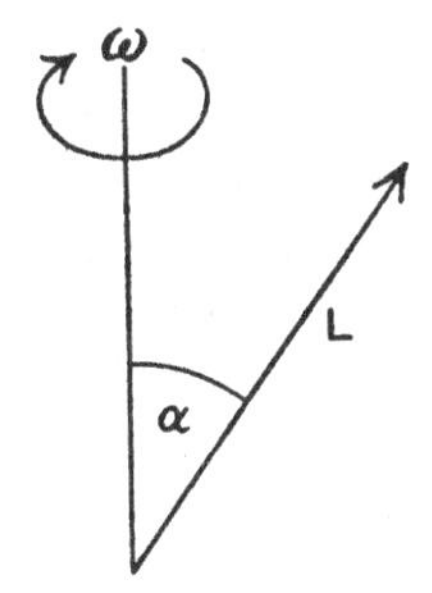

If the system of particles precesses round Oz with angular velocity ω carrying the axes Ox, Oy with it, the energy E is

$$E = \tfrac{1}{2}\Sigma\mu\,\{\dot{r}^2 + r^2\dot{\theta}^2 + r^2\sin^2\theta\,(\dot{\phi} + \omega)^2\}.$$

If ΔE is the increase of energy due to the slow precession,

$$\begin{aligned}\Delta E = E - E_0 &= \Sigma\mu r^2\sin^2\theta\dot{\phi}\omega, \text{ neglecting } \omega^2\\ &= \omega\,\{\text{angular momentum about } Oz\}\\ &= \omega L\cos\alpha, \text{ where } L \text{ is the resultant}\\ &\qquad \text{angular momentum.}\quad [\omega = 2\pi o.]\end{aligned}$$

[1] Q.T.A. § 79.

8. *Application to the normal effect.*

We had

$$\Delta E = o\,.\,mh.$$

But $$\Delta E = \omega L \cos \alpha, \text{ and } \omega = 2\pi o.$$

$$\therefore \quad mh = 2\pi L \cos \alpha,$$

and since

$$L = \frac{h}{2\pi}\sqrt{l\,(l+1)}, \quad \cos \alpha = \frac{m}{\sqrt{l\,(l+1)}}.$$

Since $-1 \leqslant \cos \alpha \leqslant 1$, m can take the values

$$l,\, l-1,\, l-2,\, \ldots,\, -(l-2),\, -(l-1),\, -l,$$

so that for a given l there are $(2l+1)$ possible m's and thus $(2l+1)$ possible Zeeman terms. These states have equal 'weight' from the point of view of statistical mechanics, so that a term whose quantum number is l has a statistical weight $(2l+1)$; but the possible combinations between such sets of terms lead only to three lines, by the correspondence principle. For example, if the atom goes from a state $l_1 = 3$ to a state $l_2 = 2$, possible m_1's are $-3, -2, -1, 0, 1, 2, 3$ and m_2's $-2, -1, 0, 1, 2$. The possible transitions are given by $|\, m_1 - m_2 \,| = 1$ or 0, and are indicated by the arrows:

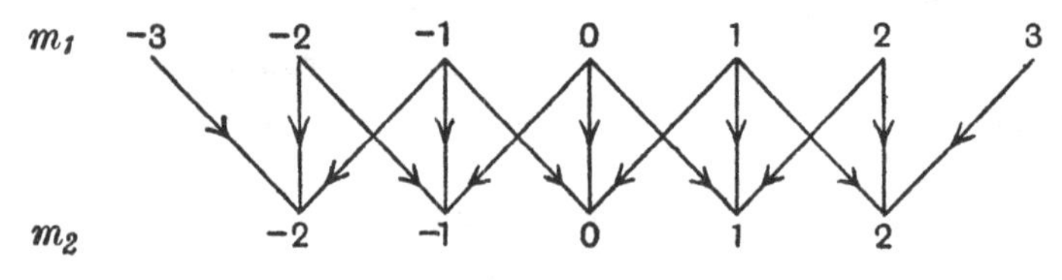

Of the 15 possible transitions, 5 have $m_1 - m_2 = 1$, 5 have $m_1 - m_2 = 0$ and 5 have $m_1 - m_2 = -1$, and each line of the Zeeman triplet is five superposed lines in this case.

9. *The anomalous Zeeman effect of the normal multiplets.*

If a multiplet is subjected to a weak magnetic field H,

each line is resolved into a large number of components whose displacements are proportional to H but different for the various members of the multiplet. If the field is gradually increased in strength so that the displacements of the components become of the same order of magnitude as the original range of the multiplet, the resolutions undergo gradual changes, until for strong fields the components of the various members of the multiplet flow together and produce a normal Zeeman triplet. This change on passing from a weak to a strong field is known as the Paschen-Back effect[1].

10. *Preston's rule.*

The Zeeman type of separation for a line given by the transition $ns \rightarrow n'p_1$ for example, depends solely upon s and p_1 and not upon n, n' the respective numbers of the terms in the s and p_1 *term* series.

Thus the Zeeman type shows clearly the series (principal, sharp, etc.) to which a line belongs.

11. *Zeeman types.*

As an illustration consider the doublet terms of an alkali given by the scheme

l \ j	$\frac{1}{2}$	$\frac{3}{2}$	$\frac{5}{2}$
0	s		
1	p_1	p_2	
2		d_1	d_2

$\left(s = \frac{1}{2}\right)$

[1] 'Zeemaneffekt und Multiplettstruktur der Spektrallinien,' by E. BACK and A. LANDÉ, Berlin, 1925, §§ 14–17.

The only possible transitions are shown by the arrows, for Δl can only be ± 1, and Δj only ± 1 or 0.

Thus the principal series multiplets are of the types sp_1, sp_2 and the diffuse series multiplets of the types $p_1 d_1$, $p_2 d_1$, $p_2 d_2$.

The Zeeman separations for these five lines are shown in the figure below:

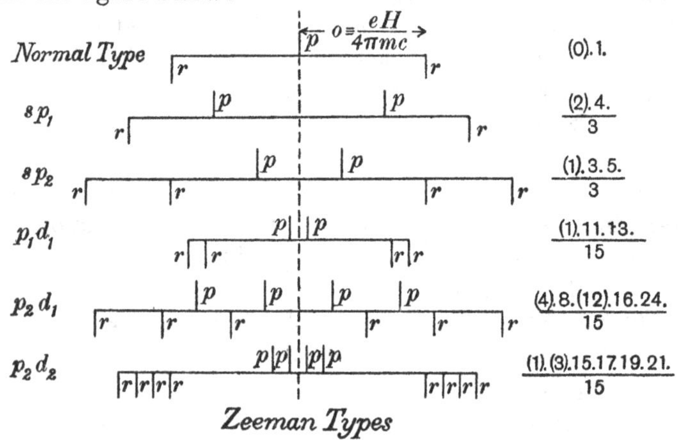

Zeeman Types

The $p_2 d_1$ type is denoted by $\dfrac{(4)\ 8\ (12)\ 16.24}{15}$; this means that the distances of the five lines on the right from the original line (whose position is the dotted central line) are $\dfrac{4}{15}, \dfrac{8}{15}, \dfrac{12}{15}, \dfrac{16}{15}, \dfrac{24}{15}$ of the normal Zeeman displacement o, which is taken as the unit of measurement. Thus if $\Delta\nu$ is the displacement, $\dfrac{\Delta\nu}{o} = \dfrac{4}{15}, \dfrac{8}{15}, \ldots$ etc., where $o = \dfrac{eH}{4\pi mc}$.

The numbers in brackets correspond to lines p polarised parallel to the field and those not in brackets to lines r polarised at right angles to the field.

The above illustrate too *Runge's rule* that the displacements are rational fractions of the normal effect ($\pm\, o$).

12. *Term analysis.*

The displaced components of a Zeeman type are now sought as the difference of two displaced terms, so that the Runge fractions become differences of two term fractions corresponding to new energy levels produced by the magnetic field.

Consider say the type $p_1 d_1$ just given, into which the multiplet $p_1 d_1$ of the diffuse series of an alkali splits up under the action of a magnetic field. The multiplet itself is due to a transition from a term d_1 to a term p_1, and it is now sought to discover new term levels into which these terms d_1 and p_1 split under the action of the magnetic field. Transitions between these two new sets of levels give rise to the Zeeman type $p_1 d_1$.

The following values for the new levels were found by Landé:

Magnetic quantum number m	$-\frac{3}{2}$	$-\frac{1}{2}$	$\frac{1}{2}$	$\frac{3}{2}$

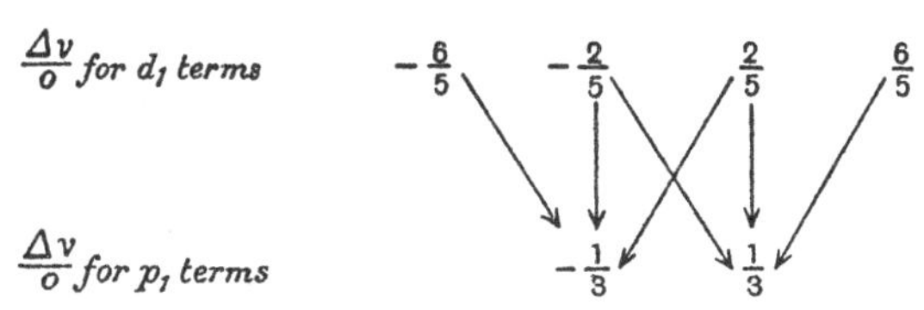

Possible transitions are shown by arrows, since Δm must be ± 1 or 0.

$\frac{\Delta \nu}{o}$ for $p_1 d_1$ has six values, viz.:

$$-\frac{6}{5}+\frac{1}{3}, \quad -\frac{2}{5}-\frac{1}{3}, \text{ where } \Delta m = +1,$$

$$\frac{6}{5}-\frac{1}{3}, \quad \frac{2}{5}+\frac{1}{3}, \text{ where } \Delta m = -1,$$

$$-\frac{2}{5}+\frac{1}{3}, \quad \frac{2}{5}-\frac{1}{3}, \text{ where } \Delta m = 0,$$

or, $\pm\frac{13}{15}$, $\pm\frac{11}{15}$, where $|\Delta m| = 1$, corresponding to r polarisation, and $\pm\frac{1}{15}$, where $\Delta m = 0$, corresponding to p polarisation, which agrees with the $\frac{(1).11.13}{15}$ of the figure of § 11, and gives the polarisations correctly.

The following figure shows the levels and the transitions:

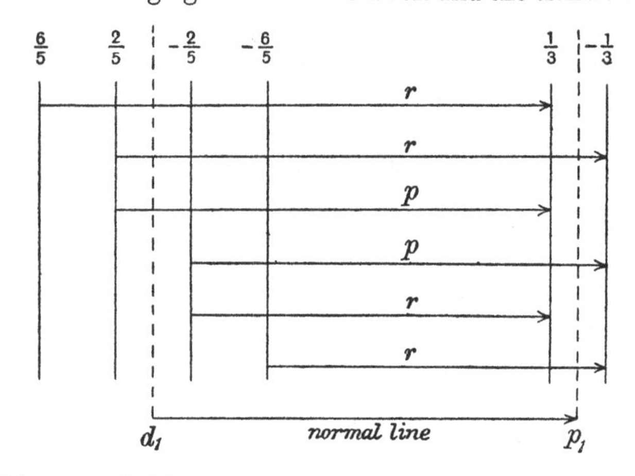

The test of this is that this scheme of Zeeman terms, as regards p_1 say, should suffice for any other line in which p_1 is concerned, as for instance the line sp_1 of the principal series.

For sp_1, we have the scheme:

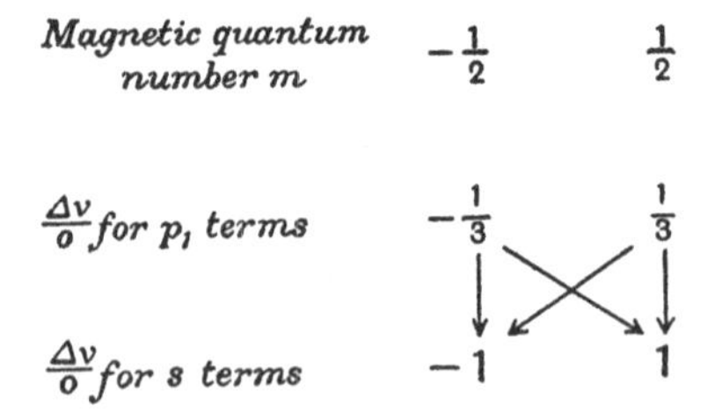

$\frac{\Delta\nu}{o}$ for sp_1 is therefore $\pm\frac{2}{3}, \pm\frac{4}{3}$, agreeing with the $\frac{(2).4}{3}$ given for sp_1 in the figure of § 11.

The term levels used for p_1 are the same as before.

Again for sp_2 we have

Magnetic quantum number m	$-\frac{3}{2}$	$-\frac{1}{2}$	$\frac{1}{2}$	$\frac{3}{2}$
$\frac{\Delta\nu}{o}$ *for s terms*		-1	1	
$\frac{\Delta\nu}{o}$ *for p_2 terms*	$-\frac{6}{3}$	$-\frac{2}{3}$	$\frac{2}{3}$	$\frac{6}{3}$

so that $\frac{\Delta\nu}{o}$ for sp_2 is $\pm\frac{1}{3}, \pm\frac{3}{3}, \pm\frac{5}{3}$, in full agreement with the $\frac{(1).3.5}{3}$ of the figure of § 11.

Then using the above results for d_1 and p_2 we have

Magnetic quantum number m	$-\frac{3}{2}$	$-\frac{1}{2}$	$\frac{1}{2}$	$\frac{3}{2}$
$\frac{\Delta\nu}{o}$ *for d_1 terms*	$-\frac{6}{5}$	$-\frac{2}{5}$	$\frac{2}{5}$	$\frac{6}{5}$
$\frac{\Delta\nu}{o}$ *for p_2 terms*	$-\frac{6}{3}$	$-\frac{2}{3}$	$\frac{2}{3}$	$\frac{6}{3}$

leading to $\frac{\Delta\nu}{o}=\frac{(4)8(12).16.24}{15}$, as in § 11.

The following is a table of the Zeeman *terms* for the doublets and triplets:

(α) *Doublets*

m	$-\frac{5}{2}$	$-\frac{3}{2}$	$-\frac{1}{2}$	$\frac{1}{2}$	$\frac{3}{2}$	$\frac{5}{2}$
s			-1	1		
p_1			$-\frac{1}{3}$	$\frac{1}{3}$		
p_2		$-\frac{6}{3}$	$-\frac{2}{3}$	$\frac{2}{3}$	$\frac{6}{3}$	
d_1		$-\frac{6}{5}$	$-\frac{2}{5}$	$\frac{2}{5}$	$\frac{6}{5}$	
d_2	$-\frac{15}{5}$	$-\frac{9}{5}$	$-\frac{3}{5}$	$\frac{3}{5}$	$\frac{9}{5}$	$\frac{15}{5}$

(β) *Triplets*

m	-3	-2	-1	0	1	2	3
s			-2	0	2		
p_1				0			
p_2			$-\frac{3}{2}$	0	$\frac{3}{2}$		
p_3		$-\frac{6}{2}$	$-\frac{3}{2}$	0	$\frac{3}{2}$	$\frac{6}{2}$	
d_1			$-\frac{1}{2}$	0	$\frac{1}{2}$		
d_2		$-\frac{14}{6}$	$-\frac{7}{6}$	0	$\frac{7}{6}$	$\frac{14}{6}$	
d_3	$-\frac{12}{3}$	$-\frac{8}{3}$	$-\frac{4}{3}$	0	$\frac{4}{3}$	$\frac{8}{3}$	$\frac{12}{3}$

13. *The Landé g-formula for the Zeeman terms.*

This is an empirical formula found by Landé which determines the Zeeman terms corresponding to any multiplet term whose quantum numbers are l, s, j. Landé showed that the Zeeman terms are given by $\dfrac{\Delta\nu}{o} = mg$, where m is $\pm j$, $\pm(j-1)$, $\pm(j-2)$, ... and g is the 'separation factor' given by the formula

$$g = 1 + \frac{j(j+1) + s(s+1) - l(l+1)}{2j(j+1)}, \text{ (\textit{the g-formula})}.$$

Thus the values of l, s, j for a multiplet term determine its Zeeman behaviour. The number of Zeeman terms into which a multiplet term splits up is $2j+1$, this being the number of possible values of m (cf. § 77).

For example, take the d_2 *doublet term*, for which

$$l = 2,\ s = \frac{1}{2},\ j = \frac{5}{2},\ (\S\ 5).$$

Then $$g = 1 + \frac{\frac{5}{2}.\frac{7}{2} + \frac{1}{2}.\frac{3}{2} - 2.3}{2.\frac{5}{2}.\frac{7}{2}} = 1 + \frac{1}{5} = \frac{6}{5},$$

and $\dfrac{\Delta\nu}{o} = mg$, where m is $\pm\frac{5}{2}, \pm\frac{3}{2}, \pm\frac{1}{2}$.

$$\therefore\ \frac{\Delta\nu}{o} = \pm\frac{5}{2}.\frac{6}{5},\ \pm\frac{3}{2}.\frac{6}{5},\ \pm\frac{1}{2}.\frac{6}{5}$$

$$= \pm\frac{15}{5},\ \pm\frac{9}{5},\ \pm\frac{3}{5},$$

agreeing with the value given in table (α) of § 12.

Again for the d_2 *triplet term*, $l = 2$, $s = 1$, $j = 2$, so that

$$g = 1 + \frac{2.3 + 1.2 - 2.3}{2.2.3} = 1 + \frac{1}{6} = \frac{7}{6},$$

and $\dfrac{\Delta\nu}{o} = mg$, where $m = \pm 2, \pm 1, 0$, so that

$$\frac{\Delta\nu}{o} = \pm 2.\frac{7}{6},\ \pm 1.\frac{7}{6},\ 0.\frac{7}{6}$$

$$= \pm\frac{14}{6},\ \pm\frac{7}{6},\ 0,$$

as given in table (β) of § 12.

The following is a table of g values:

Term	*Singulets*			*Doublets*						*Triplets*							
	s	p	d	s	p_1	p_2	d_1	d_2	...	s	p_1	p_2	p_3	d_1	d_2	d_3	...
g	1	1	1	2	$\frac{2}{3}$	$\frac{4}{3}$	$\frac{4}{5}$	$\frac{6}{5}$	...	2	$\frac{3}{2}$	$\frac{3}{2}$	$\frac{3}{2}$	$\frac{1}{2}$	$\frac{7}{6}$	$\frac{4}{3}$	...

CHAPTER IV

ATOMIC MAGNETISM; THE BOHR MAGNETON; THE STERN-GERLACH EXPERIMENT; MAGNETISM AND TEMPERATURE; THE MAGNETO-MECHANICAL EFFECT

14. *Atomic magnetism.*

An atom, with its revolving electrons, has magnetic properties revealed by its Zeeman effect and different for the different stationary states.

If M is the magnetic moment, supposed in the direction of the resultant angular momentum J, the change of energy due to the action of a magnetic field H is $\Delta E = HM \cos(HM)$, where (HM) is the angle between H and M. But this $\Delta E = h\Delta\nu$, where $\Delta\nu$ is the Zeeman displacement.

$$\therefore \quad h\Delta\nu = HM \cos(HM).$$

But from § 7,

$$h\Delta\nu = \Delta E = \omega J \cos(JH) = 2\pi o . J \cos(JH)$$

in the *normal* Zeeman effect.

Hence, since $(HM) = (JH)$, we have

$$HM = 2\pi o . J = 2\pi \left(\frac{eH}{4\pi m_0 c}\right) J,$$

where now m_0 is the mass of the electron (as from now m is needed to denote a quantum number);

$$\therefore \quad \frac{M}{J} = \frac{e}{2m_0 c},$$

or the ratio of the magnetic moment to the mechanical moment is $e/2m_0c$, for atoms which show the normal Zeeman effect.

15. *The Bohr magneton.*

If the unit of angular momentum is $\frac{h}{2\pi}$, the corresponding unit of magnetic moment is $\frac{e}{2m_0c}\cdot\frac{h}{2\pi}=\frac{eh}{4\pi m_0c}$; this unit of magnetic moment is the Bohr magneton. Using the known values of e, h, m_0, c its value is $\cdot 921\times 10^{-20}$ ergs/gauss.

Thus the angular momentum measured in units $\frac{h}{2\pi}$ is equal to the magnetic moment measured in magnetons, for an atom whose Zeeman effect is normal.

16. *The field components of M and J.*

If M_H is the atomic magnetic moment resolved along the magnetic field H, and J_H the corresponding component of the mechanical moment, then

$$\frac{M_H}{J_H}=\frac{M}{J}=\frac{e}{2m_0c}.$$

Now $J_H=\frac{mh}{2\pi}$, where m is the quantum number used in the theory of the Zeeman effect;

$$\therefore\ M_H=m\left(\frac{eh}{4\pi m_0c}\right).$$

Thus m denotes the angular momentum of the atom about H in units $\frac{h}{2\pi}$ and also the magnetic moment of the atom along H in magnetons, for an atom with the normal Zeeman effect.

But for an atom with an anomalous Zeeman effect, o is replaced by go, where g is the separation factor (so that the Larmor precession is go in these states); hence

$$\Delta E = HM\cos(HM) = HM_H$$

and

$$\Delta E = 2\pi go\, J\cos(JH) = 2\pi go\, J_H;$$

$$\therefore\ \frac{M_H}{J_H}=\frac{2\pi go}{H}=g\cdot\frac{e}{2m_0c}.$$

But $$J_H = \frac{mh}{2\pi}.$$

$$\therefore\ M_H = gm\left(\frac{eh}{4\pi m_0 c}\right),$$

so that if m denotes the angular momentum about H in units $\frac{h}{2\pi}$, mg denotes the magnetic moment along H in magnetons.

Thus for example in the term p_2 of the doublets

$$m = \frac{3}{2} \quad \frac{1}{2} \quad -\frac{1}{2} \quad -\frac{3}{2}, \qquad \left(g = \frac{4}{3}\right),$$

$$\frac{\Delta\nu}{o} = mg = \frac{6}{3} \quad \frac{2}{3} \quad -\frac{2}{3} \quad -\frac{6}{3},$$

so that the first row gives the mechanical component in units $h/2\pi$ and the second the magnetic component in magnetons.

17. *The experiments of Stern and Gerlach on the magnetic deviation of silver rays*[1].

These experiments lead to a direct measurement of atomic magnetic moment components (M_H). Stern and Gerlach passed the atoms of silver vapour through a *non-uniform* magnetic field and measured the deviation from their otherwise rectilinear paths by intercepting them on a screen. The displacement thus measured determines M_H. The theory of the experiment is as follows:

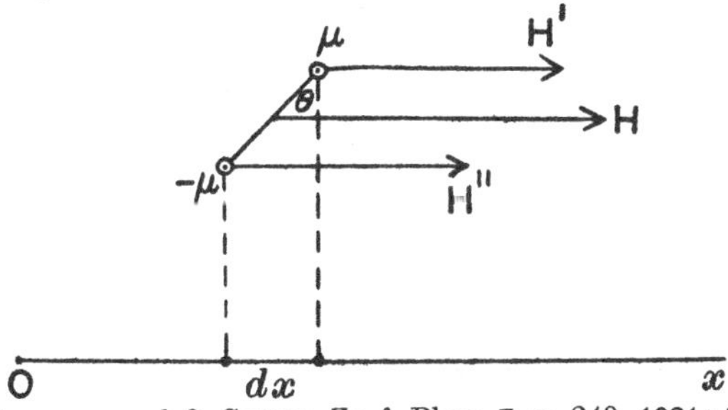

[1] W. Gerlach and O. Stern, Zs. f. Phys. 7, p. 249, 1921; 8, p. 110, 1921; 9, p. 349, 1922.

The figure shows a magnetic element (poles $\pm\mu$ at a distance b apart) in a variable field H; the axis of x is taken parallel to the field. The translatory force on the magnet is $H'\mu + H''(-\mu)$, where H', H'' are the values of H at the poles. Thus the force is

$$(H' - H'')\,\mu = \frac{dH}{dx}\,.\,dx\,.\,\mu = \frac{dH}{dx}\,(b\cos\theta)\,\mu$$

$$= \frac{dH}{dx}\,M\cos\theta, \text{ where } M \text{ is the moment,}$$

$$= M_H\frac{dH}{dx}$$

$= M_H G$, where G is the known *gradient* of the magnetic field.

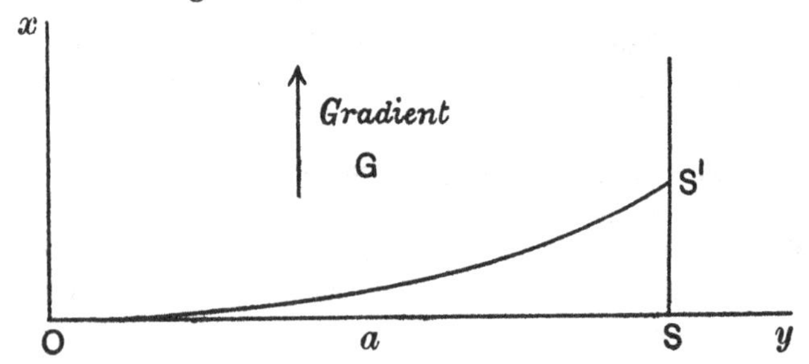

For a silver atom of mass m_s,

$$\left.\begin{aligned} m_s\ddot{x} &= M_H.G \\ m_s\ddot{y} &= O \end{aligned}\right\}; \quad \therefore \quad \begin{aligned} m_s\dot{x} &= M_H G t \\ \dot{y} &= v \end{aligned},$$

where v is the velocity at O;

$$\therefore \left.\begin{aligned} m_s x &= \frac{1}{2}M_H G t^2 \\ y &= vt \end{aligned}\right\}.$$

$\therefore\ x = \dfrac{M_H G y^2}{2m_s v^2}$; and if a is the distance of the screen from O, the deviation SS' on the screen is $M_H G a^2/2m_s v^2$, and thus M_H is found.

In their first paper, using silver atoms, there were two sharply defined positions of S'. Exact measurements showed that M_H was $\pm$ 1 magneton. But for silver in the normal state (which for it is the s term of the doublets)

$$\frac{\Delta\nu}{o} = mg = \pm 1, \text{ for } m = \frac{1}{2}, -\frac{1}{2}$$

$$g = 2\,, \qquad (\S\ 13)$$

so that $$mg = 1, -1.$$

This result, with others, is a striking confirmation of the truth of Landé's theory of the Zeeman states, and indeed is a direct proof of the existence of the 'term states' of the Bohr theory; up to this time these states were no more than a mental picture of something whose *differences* gave the spectral lines themselves.

It is now apparent that we may use the mg values of the Zeeman tables as predictions of the Stern-Gerlach deviations in magneton units for different atoms. Hence the table:

Normal state	*mg*	*Stern-Gerlach deviation*
A singulet s term	0	
doublet s term	$-1 \quad 1$	
doublet p_1 term	$-\frac{1}{3} \quad \frac{1}{3}$	
doublet p_2 term	$-\frac{6}{3} \quad -\frac{2}{3} \quad \frac{2}{3} \quad \frac{6}{3}$	
triplet s term	$-2 \quad 0 \quad 2$	
quartet s term	$-3 \quad -1 \quad 1 \quad 3$	

The predictions of the Zeeman terms have always been exactly fulfilled in these experiments; Ag, Cu, Au show deviations $\pm$ 1, Pb deviation 0, Tl $\pm \frac{1}{3}$, this last showing that the normal state is a doublet p_1 term (Grotrian)[1].

[1] W. Grotrian, Zs. f. Phys. 12, p. 218, 1923.

18. *Magnetism and temperature.*

The magnetic energy of an atom $E = -HM_H$, in the notation of § 14. If dW is the probability that the atom has energy E between $-HM_H$ and $-H(M_H + dM_H)$, $dW = Ae^{-E/kT}\,dM_H$, where T is the temperature, k the Boltzmann constant and A is a constant.

$$\therefore\ dW = Ae^{HM_H/kT}\,dM_H$$
$$= A\left(1 + \frac{HM_H}{kT}\right) dM_H,$$

approx., expanding the exponential, the term $\frac{HM_H}{kT}$ being small and so H not too large.

In the absence of H the probability is 1, so that

$$\int dW = 1; \text{ or } \int_{-M}^{M} A\,dM_H = 1; \text{ or } A = \frac{1}{2M};$$
$$\therefore\ dW = \frac{1}{2M}\left(1 + \frac{HM_H}{kT}\right) dM_H.$$

The magnetic moment m_H per mol is

$$\int_{-M}^{M} NM_H\,dW = \int_{-M}^{M} \frac{N}{2M} M_H \left(1 + \frac{HNM_H}{RT}\right) dM_H,$$

where N is the number of atoms per mol, and the mol gas constant $R = Nk$,

$$= \frac{N\int_{-M}^{M} M_H\,dM_H + \frac{HN^2}{RT}\int_{-M}^{M} M_H^2\,dM_H}{2M}.$$
$$\therefore\ m_H = N\overline{M_H} + \frac{HN^2}{RT}\overline{M_H^2},$$

where the bars denote mean values for the different orientations of the magnetic axes of the atoms. This is Langevin's formula for the component magnetic moment m_H per mol of the gas parallel to the inducing field H.

The divergence between classical and quantum theory arises in the calculation of these mean values; whereas on the classical theory all orientations are equally probable, on the quantum theory only certain ones are possible, as the experiments of Stern and Gerlach have shown. On both theories $\overline{M_H} = 0$; because for the classical theory

$$\overline{M_H} = \frac{1}{2M}\int_{-M}^{M} M_H dM_H = 0,$$

and for the quantum theory the possible directions of the atomic axes produce values of M_H equal and opposite in pairs, so that $\Sigma M_H = 0$.

Thus on both theories $m_H = \dfrac{HN^2\overline{M_H^2}}{RT}$.

On the classical theory

$$\overline{M_H^2} = \frac{1}{2M}\int_{-M}^{M} M_H^2 dM_H = \frac{1}{3}M^2,$$

so that

$$m_H = \frac{H}{3RT}M^2N^2,$$

$$\therefore M = \frac{1}{N}\sqrt{\frac{3RTm_H}{H}},$$

the classical result of Langevin.

Weiss[1] tested this by measurements of m_H, H, T which gave M values which were not multiples of Bohr's magneton, but of a unit about five times smaller (the Weiss magneton). This result was for some time a serious difficulty for the quantum theory.

Pauli[2] however pointed out that from the quantum theory standpoint the calculation of $\overline{M_H^2}$ must be modified, as mentioned above. For on this theory (§ 16)

$$M_H = mg \text{ magnetons} = mg\,\square, \text{ where } \square \equiv \frac{eh}{4\pi m_0 c}.$$

[1] P. Weiss, Phys. Zs. **12**, p. 935, **1911.**
[2] W. Pauli, Phys. Zs. **21**, p. 615, **1920.**

The possible values of m are

$$j, j-1, j-2, \ldots -(j-1), -j,$$

and therefore of M_H

$$jg\,\square,\ (j-1)\,g\,\square, \text{ etc.}$$

$$\therefore\ \overline{M_H^2} = \frac{g^2\,\square^2\,[j^2+(j-1)^2 \ldots + (j-1)^2 + j^2]}{2j+1}$$

$$= g^2\,\square^2\,\frac{2j\,(j+1)\,(2j+1)}{6}\cdot\frac{1}{2j+1} = g^2\,\square^2\,\frac{(j+1)\,j}{3},$$

$$\therefore\ m_H = \frac{HN^2\,\overline{M_H^2}}{RT} = \frac{H}{RT}\,N^2 g^2\,\square^2\,\frac{j\,(j+1)}{3}.$$

If S is the mol magnetic susceptibility, S is the value of m_H when $H = 1$,

so that $$ST = \frac{N^2 g^2\,\square^2}{R}\cdot\frac{j\,(j+1)}{3}.$$

$$\therefore\ g\sqrt{j\,(j+1)} = \frac{\sqrt{3RST}}{N\square}$$

$$= \frac{\sqrt{3RST}\,4\pi m_0 c}{Neh};$$

$$\therefore\ g\sqrt{j\,(j+1)} = 2{\cdot}83\sqrt{ST},$$

using the well-known values of the constants R, m_0, c, N, e, h.

This formula has been tested by the experiments of Cabrera[1] and St Meyer[2] on the triply ionised rare earths from Ce_{+++} to Yb_{+++}; $2{\cdot}83\sqrt{ST}$ was found as the result of experiment and $g\sqrt{j\,(j+1)}$ calculated from the theoretical values of j and g; the agreement was close. Thus the smaller Weiss unit of the classical theory is not required; the Bohr unit suffices.

[1] B. Cabrera, Comptes Rendus, **180**, p. 668, **1925**.
[2] St Meyer, Phys. Zs. **26**, pp. 1 and 478, **1925**.

19. *The Richardson-Barnett magneto-mechanical effect*[1, 2]. If a diamagnetic body is set in rotation with angular velocity ω about an axis, the atom axes are made to undergo a precession ω and the body becomes a magnet whose field is H; in order to preserve the direction of their angular momentum in space, the electron orbits within the atom make a counter precession in the opposite sense at a rate go (§ 16), where $o = \frac{eH}{4\pi m_0 c}$ and g is the Landé factor. This go must be equal to ω, so that

$$\omega = \frac{geH}{4\pi m_0 c} \quad \text{or } H = \frac{4\pi m_0 c\omega}{ge} \quad \text{...........(1).}$$

Assuming the normal Larmor precession, $g = 1$ and H should be $\frac{4\pi m_0 c\omega}{e}$. The actual value found was always one-half of this, and before the Landé theory was supposed to be due to an anomalous doubling of the Kaufmann ratio of e/m_0; after the Landé theory the halved value meant merely that $g = 2$ in the formula (1), and this is the value of g for a ground state of the s type in Zeeman theory (cf. table of § 13).

The experiments of Einstein and de Haas[3], and later ones of greater precision by Back[4], confirm this view. A revolving body was suddenly magnetised and the ratio of the mechanical moment to the magnetic moment produced determined. On the old theory the normal value $e/2m_0 c$ was expected, but the experiments gave double this value, accounted for by the precession go, where $g = 2$.

[1] O. W. Richardson, Phys. Rev. **26**, p. 248, **1908**.
[2] J. S. Barnett, Phys. Rev. **6**, p. 239, **1915**.
[3] A. Einstein and de Haas, Verh. Deutsch. Phys. Ges. **17**, p. 152, **1915**.
[4] E. Back, Ann. der Phys. **60**, p. 109, **1919**.

CHAPTER V

INTERPRETATION OF THE g-FORMULA; THE PAULI VERBOT; THE SPINNING ELECTRON; SPIN DOUBLETS

20. *The interpretation of the g-formula.*

This remarkable empirical formula, which clears up the main complexities of the anomalous Zeeman effect, can be given the following interpretation on the Landé model:

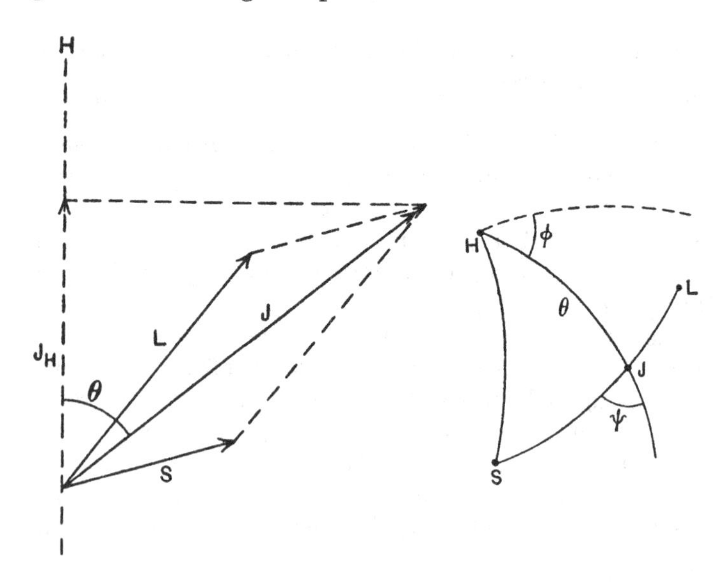

In the first figure, LJS is the Landé model of a multiplet term (l, j, s) in which the parallelogram formed by LS precesses round its diagonal J owing to the 'coupling' forces between the core and the series electron. Under the action of a magnetic field H (not so strong as to break down this coupling action) J precesses round H. The second figure represents the surface of a sphere through whose centre are drawn lines parallel to S, J, L, H of the first figure meeting the surface in the points S, J, L, H. The dotted arc is some fixed arc through H. The two precessions round J and H are given by the rates of increase of ψ and ϕ.

As before J, L, S are respectively

$$\frac{h}{2\pi}\sqrt{j(j+1)},\quad \frac{h}{2\pi}\sqrt{l(l+1)},\quad \frac{h}{2\pi}\sqrt{s(s+1)},$$

and J_H, the projection of J upon H, $= \dfrac{mh}{2\pi}$.

The g-formula is

$$g = 1 + \frac{j(j+1) + s(s+1) - l(l+1)}{2j(j+1)}$$

$$= 1 + \frac{J^2 + S^2 - L^2}{2J^2}.$$

But $$\cos(JS) = \frac{J^2 + S^2 - L^2}{2JS},$$

so that $$g = 1 + \frac{S}{J}\cos(JS).$$

If $$\theta = (JH), \quad \cos\theta = \frac{J_H}{J} = \frac{mh}{2\pi J},$$

$$\therefore\ m = \frac{2\pi J\cos\theta}{h}.$$

$$\therefore\ mg = \frac{2\pi}{h}\cos\theta\{J + S\cos(JS)\}.$$

Now $\Delta\nu$, the Zeeman separation, $= mgo$,

$$\therefore\ \frac{h}{2\pi}\frac{\Delta\nu}{o} = J\cos\theta + S\cos\theta\cos(JS)\ \ldots\ldots(1).$$

From the spherical triangle of the second figure,

$$\cos(SH) = \cos\theta\cos(JS) - \sin\theta\sin(JS)\cos\psi.$$

Taking time mean values, denoted by a bar,

$$\overline{\cos(SH)} = \cos\theta \cos(JS) \quad \text{...........(2)},$$

since $\overline{\cos\psi} = 0$ owing to the periodicity of $\cos\psi$.

Also $J\cos\theta = J_H = L\cos(LH) + S\cos(SH)$,

whence

$$J\cos\theta = L\overline{\cos(LH)} + S\overline{\cos(SH)}, \text{ taking mean values...(3).}$$

Using (2) and (3) in (1) we have

$$\frac{h\Delta\nu}{2\pi o} = L\overline{\cos(LH)} + 2S\overline{\cos(SH)},$$

$$h\Delta\nu = \frac{eH}{2m_0c}[L\overline{\cos(LH)} + 2S\overline{\cos(SH)}], \quad \text{...(4)},$$

inserting the value of o.

Thus the series electron has additional energy due to the field equal to that of a magnet of moment $\frac{e}{2m_0c}L$, and the core that of a magnet of moment $2.\frac{e}{2m_0c}S$. For the series electron the ratio of magnetic to mechanical moment has the classical value $e/2m_0c$, while for the core the ratio is twice as large, namely e/m_0c.

These assumptions are necessary to account for the g-formula by the Landé model.

Pauli's criticism of the core theory.

In December 1924, Pauli[1] showed that in the case of the alkalis of higher atomic number (Z) where the K electrons are revolving near a large nuclear charge Ze, a pronounced relativity mass variation must occur, so that the ratio of magnetic to mechanical moment ($= e/m_0c$) for the core must show marked change with Z, on account of the relativity variation of m_0; the value of m_0 would

[1] W. Pauli, Zs. f. Phys. **31**, p. 373, **1925**.

depend upon the orbital velocity of these K electrons of the core, and this in turn upon Z. This means that the nature of the Zeeman effect must depend upon nuclear charge, if the core is the seat of the mechanical moment S, or the magnetic moment eS/m_0c.

Pauli showed that the g-formula would become

$$g = 1 + \gamma \left\{ \frac{j(j+1) + s(s+1) - l(l+1)}{2j(j+1)} \right\},$$

where $\gamma = 1 - \frac{\alpha^2}{2}(Z-1)^2$, and $\alpha = \frac{2\pi e}{hc}$, the fine structure constant[1] of Sommerfeld. ($\alpha^2 = 5{\cdot}31 \times 10^{-5}$.)

Thus for Na ($Z = 11$), $\gamma = 1 - {\cdot}003$,
K ($Z = 19$), $\gamma = 1 - {\cdot}009$,
Rb ($Z = 37$), $\gamma = 1 - {\cdot}034$,
Cs ($Z = 55$), $\gamma = 1 - {\cdot}078$.

Experiment shows that there is no dependence, of this kind, of the Zeeman effect upon nuclear charge, so that it must be inferred that the core of the alkalis (which is the same as the atom of the inert gas just preceding the alkali in the periodic table) has no mechanical or magnetic moment.

[On the Landé core theory, the core electrons of the alkalis would have angular momentum corresponding to $s = \frac{1}{2}$ and therefore equal to

$$\frac{h}{2\pi}\sqrt{s(s+1)} \text{ or } \frac{h}{2\pi}\sqrt{\frac{3}{4}}.]$$

21. *Pauli's exclusion principle.* (*The Pauli verbot.*)

A month after the criticism of § 20, Pauli had turned to a new line of thought, hinted at in that paper; and this was that the quantum number s must be associated with the series electron and dissociated from the core. Thus all the quantum numbers were assigned to the series electron

[1] Q.T.A. § 60.

alone, though nothing was said as to what the electron was to do with the corresponding momentum S.

Pauli[1] then formulated the following axiom: 'Every electron in an atom has four quantum numbers (n, l, j, m) and no two electrons in the atom can have the same four quantum numbers.'

This is Pauli's exclusion principle or the Pauli 'verbot,' which leads in a striking manner to a theory of the possible number of electrons, 2, 8, 18, 32, ... in the $K, L, M, N, \ldots$ shells of the elements in the periodic table (§ 29).

22. *The spinning electron.*

In the above theory, Pauli is not committed to any form of model representation of the atom; the quantum number s has lost its original Landé significance and has been transferred from the core to the electron which has now four degrees of quantum freedom.

In November 1925, Goudsmit and Uhlenbeck[2] proposed their theory of the spinning electron; the electron was supposed to be spinning about an axis like a top or a planet, and it was suggested that the angular momentum S previously assigned to the core was in fact the angular momentum of the electron about its axis.

The idea of a quantised spinning electron had been used by Compton[3] in 1921 in relation to the magneton, but the thought of using it to clear up this Landé-Pauli difficulty (§§ 20, 21) originated with Goudsmit and Uhlenbeck, who were then unaware of Compton's idea.

It is surprising that while nearly all the attributes of an astronomical model, the solar system (sun, planets and their orbits), had been taken over into the theory of the

[1] W. Pauli, Zs. f. Phys. **31**, p. 765, Jan. **1925**.

[2] G. E. Uhlenbeck and S. Goudsmit, Naturwissensch. **47**, p. 953, Nov. **1925**; Nature, **117**, p. 264, Feb. **1926**.

[3] A. H. Compton, Jour. Franklin Inst. **192**, p. 145, **1921**.

atom, the important feature of the rotation of the planets about their axes had escaped attention until then.

The quantum numbers have now the following meaning; as before, n, l are the principal and azimuthal quantum numbers of the electron in its path and m is the azimuthal quantum number representing its potential behaviour in a magnetic field, but now s is the quantum number for the spin of the electron itself.

The anomaly of § 20, that the magnetic moment corresponding to S was eS/m_0c, just twice the normal value, now disappears. For it is found that Abraham's old calculation[1] of the ratio of magnetic to mechanical moment for a spinning charged sphere gives the result e/m_0c. This is twice the normal value for an electron revolving in an orbit, which is just what is needed in the g-formula theory. Abraham found for a spinning charged sphere that the rotational energy T

$$= \frac{1}{9} \frac{e^2 a}{c^2} \dot{\phi}^2,$$

where a is the radius of the sphere. Therefore the mechanical moment p_ϕ

$$= \frac{\partial T}{\partial \dot{\phi}} = \frac{2}{9} \frac{e^2 a}{c^2} \dot{\phi}.$$

Also the magnetic moment was found to be $\frac{1}{3} \frac{ea^2}{c} \dot{\phi}$ and the mass m_0 to be $\frac{2}{3} \frac{e^2}{c^2 a}$, so that the ratio of magnetic to mechanical moment $= \frac{3ac}{2e} = \frac{e}{m_0 c}$.

23. *Spin doublets.*

Consider a stationary state in which an electron describes an orbit about a core producing a central field. The orbit, if the spin of the electron is neglected, is of the rosette

[1] M. Abraham, Ann. der Phys. **10**, p. 105, **1903**.

type[1], where there is a secular precession of the orbital form in its own plane; the quantum numbers are n, l. (The k of Sommerfeld's rosette theory is our $l+1$.)

The electron, owing to its velocity $\mathbf{v}$ in the central field $\mathbf{E}$, is acted upon by a magnetic field $\frac{1}{c}(\mathbf{Ev})$, which is perpendicular to the plane of the orbit; owing to its spin the electron is a small magnetic top upon which this magnetic field produces a couple whose axis lies in the plane of the orbit. This causes a precession of the spin axis about the normal to the plane of the orbit at such a rate as to compensate for the orbital precession and so conserve the angular momentum of the atom. Thus corresponding to the state n, l with no electron spin, there will in general exist a set of states which differ in the inclination of the spin axis to the orbital plane defined by the quantum number s. If the spin corresponds to a one quantum rotation, $s=\frac{1}{2}$ and there are two such states; this agrees with the value used in Zeeman theory for the alkalis, with their *single* series electron.

Then since (§ 4)

$$|l-s| \leqslant j \leqslant l+s,$$

we have

$$\left|l-\frac{1}{2}\right| \leqslant j \leqslant l+\frac{1}{2},$$

so that $j=l\pm\frac{1}{2}$, except when $l=0$ when $j=\frac{1}{2}$ only. It is at once apparent by reference to the table of the doublets, § 11, that the two values of j, namely $l-\frac{1}{2}$ and $l+\frac{1}{2}$, correspond to the doublet separation for a given l; thus for $l=1$, $j=\frac{1}{2}$ and $\frac{3}{2}$ give the separation of the doublet

[1] Q.T.A. § 67 and figure of § 64.

terms p_1, p_2; for $l = 2, j = \frac{3}{2}$ and $\frac{5}{2}$ give the separation of the doublet terms d_1, d_2, and so on. The two values $j = l \pm \frac{1}{2}$ depend upon whether the direction of the spin is with or against the direction of the orbital precession; the energy difference between the states n, l, j, where $j = l \pm \frac{1}{2}$, is the 'spin doublet' separation.

24. *Early difficulties of the spin doublet theory.*

In November, just after the publication of the Naturwissenschaften[1] note by Goudsmit and Uhlenbeck, Heisenberg and Pauli made a preliminary calculation on the older quantum theory which led to a spin doublet separation twice as large as the experimentally observed values. It was hoped that the calculations made by the new mechanics might put this right; at this stage, Goudsmit and Uhlenbeck, encouraged by Bohr (who had visited Leiden just before Christmas and had taken up the idea with great enthusiasm) wrote the note in Nature[2] on the spin theory of the doublets, which appeared with the mystery of the factor 2 still unsolved. Just after this note had been written, Pauli found that the new mechanics did not do away with the factor 2, but almost simultaneously Thomas[3] showed that the discrepancy disappeared when the kinematical problem is examined more closely from the point of view of the theory of relativity.

The precession of the spin axis previously calculated was that in a system of coordinates in which the centre of the electron is at rest, whereas the nucleus is at rest and the electron is in motion. Lorentz transformations from the electron to the nucleus sufficed (§ 120).

[1,2] Foot of p. 38.
[3] L. H. THOMAS, Nature, 117, p. 514, April 1926.

CHAPTER VI

X-RAY SPECTRA AND THEIR MULTIPLET THEORY ON THE NEW MECHANICS; SCREENING AND SPIN DOUBLETS

25. *X-ray spectra and the spinning electron.*

The theory of the spinning electron has led to a new insight into the meaning of the multiple structure of X-ray spectra and has cleared up the theoretical difficulties which were met with in the attempts to account for the so-called *screening* and *relativity* doublets of these spectra by relativity considerations alone; it promises also to throw light on the so-called 'branching' of spectra, which usually accompanies the addition of a further electron to an atom and of which so far no satisfactory explanation has been given.

26. *X-ray spectra (Kossel's theory).*

The electrons revolving round the nucleus form the K, L, M, ... shells reckoning outwards from the nucleus, the corresponding principal quantum numbers being $n = 1, 2, 3, \ldots$.

A K line of the X-ray spectrum is produced by the ejection of a K electron from the atom and its replacement by an electron from one of the outer shells $L, M, \ldots$, which transition causes the radiation; an L line is due to the ejection of an L electron and its replacement by an electron from one of the outer shells $M, N, \ldots$; and so on. The K lines tend to a limit corresponding to the fall of an electron from the edge of the atom to a K level. This limit is found by observation of the 'absorption edge' of the K lines, and if ν is the frequency corresponding to this 'edge,' $h\nu$ is the energy 'term' of the K group[1].

27. *Screening number.*

The action of the nucleus upon an electron in an L orbit, for instance, is affected (i) by the inner K electrons which to a first approximation behave as if at the nucleus, (ii) by the outer M, N electrons which behave approximately like the charge on a spherical shell and thus their

[1] Q.T.A. chaps. XIV, XV.

effect is small, (iii) by the other L electrons, which form a central repulsive force. Thus on the whole the electron is acted upon by a central force due to an effective nuclear charge $(Z - \sigma)\,e$, where σ is a 'screening number' representing the effect of (i), (ii), (iii).

The experiments of Moseley and Darwin can be explained by supposing the X-ray 'terms' of an element whose atomic number is Z to be given by

$$N = (Z - \sigma_1)^2 R/1^2, \text{ for the } K \text{ terms,}$$
$$N = (Z - \sigma_2)^2 R/2^2, \text{ for the } L \text{ terms,}$$
$$N = (Z - \sigma_3)^2 R/3^2, \text{ for the } M \text{ terms,}$$

and so on, where $\sigma_1 < \sigma_2 < \sigma_3 \ldots$, N is the wave number and R is the Rydberg constant 109677.

28. *Bohr's grouping of the electrons into shells*[1].

Z	Element \ n	1 K	2 L	3 M	4 N	5 O	6 P	7 Q	
2	He	2							
4	Be	2	2						
10	Ne	2	8						
12	Mg	2	8	2					
18	A	2	8	8					
20	Ca	2	8	8	2				*iron group*
				←					
30	Zn	2	8	18	2				
36	Kr	2	8	18	8				
38	Sr	2	8	18	8	2,			*palladium group*
					←				
48	Cd	2	8	18	18	2			
54	X	2	8	18	18	8			
56	Ba	2	8	18	18	8	2		*rare earths*
					←				
80	Hg	2	8	18	32	18	2		
86	Nt	2	8	18	32	18	8		
88	Ra	2	8	18	32	18	8	2	
118	?	2	8	18	32	32	18	8	

[1] Q.T.A. chap. XVI.

The preceding table shows the building of atoms with increase of atomic number; the lowest member of each set is an inert gas He, Ne, A, Kr, X or Nt, and in each of these the shells involved are *just* complete; the M shell is completed with A, the N shell with Kr, and so on. Each inert gas is followed next in the periodic table by an alkali, so that the core of an alkali emitting an arc spectrum is one of the inert gas atoms, which as we have seen (§ 21), has no mechanical or magnetic moment.

29. *Deduction of the Bohr groups by the Pauli verbot.* The four quantum numbers of Pauli (n, l, j, m) for an electron satisfy the conditions

$$n = 1, 2, 3, \ldots,$$

$$l = 0, 1, 2, \ldots (n-1).$$

$$j = l \pm \frac{1}{2}, \text{ except for } l = 0, \text{ where } j = \frac{1}{2} \text{ only (§ 23)},$$

$$m = \pm j, \pm (j-1), \ldots \text{(§ 13)}.$$

Thus for the K shell,

$$n = 1, \quad l = 0, \quad j = \frac{1}{2}, \quad m = \pm \frac{1}{2},$$

so that only 2 electrons can be in the shell, by the verbot principle, § 22.

For the L shell,

$$n = 2, \begin{cases} l = 0, & j = \frac{1}{2}, \quad m = \pm \frac{1}{2}, \\ l = 1, & \begin{cases} j = \frac{1}{2}, & m = \pm \frac{1}{2}, \\ j = \frac{3}{2}, & m = \pm \frac{3}{2}, \quad \pm \frac{1}{2}, \end{cases} \end{cases}$$

so that 8 electrons form the full shell.

For the M shell,

$$n = 3, \quad \begin{cases} l = 0, & j = \frac{1}{2}, \quad m = \pm \frac{1}{2}, \\ l = 1, & \begin{cases} j = \frac{1}{2}, & m = \pm \frac{1}{2}, \\ j = \frac{3}{2}, & m = \pm \frac{3}{2}, \pm \frac{1}{2}, \end{cases} \\ l = 2, & \begin{cases} j = \frac{3}{2}, & m = \pm \frac{3}{2}, \pm \frac{1}{2}, \\ j = \frac{5}{2}, & m = \pm \frac{5}{2}, \pm \frac{3}{2}, \pm \frac{1}{2}, \end{cases} \end{cases}$$

so that 18 electrons form the full shell.

For the N shell this scheme requires 32 electrons, and so on, the whole being in agreement with Bohr's numbers derived from very general considerations of spectra (§ 28).

30. *The Bohr-Coster diagram*[1].

In this diagram, p. 46, $\sqrt{\frac{N}{R}}$ is plotted against Z for the K, L, M, ... terms of the elements. The formula

$$\sqrt{\frac{N}{R}} = \frac{Z - \sigma_1}{1}$$

shows that for the K terms the graph is a straight line, if the σ_1 remains constant for all the elements; the same applies to the L, M, ... terms. On looking at the Bohr table (§ 28) it is seen that the full M shell of 18 electrons is not complete until $Z = 30$, so that the screening number σ_3 is not constant until $Z = 30$; thus the graph in the Bohr-Coster diagram of the M terms up to $Z = 30$ is irregular and after that remains straight.

Again, the N shell is changing up to $Z = 48$ when it has 18 electrons; no further change occurs until $Z = 56$, when changes begin, until at $Z = 70$, the N shell has its full

[1] N. Bohr and D. Coster, Zs. f. Phys. 12, p. 342, 1923.

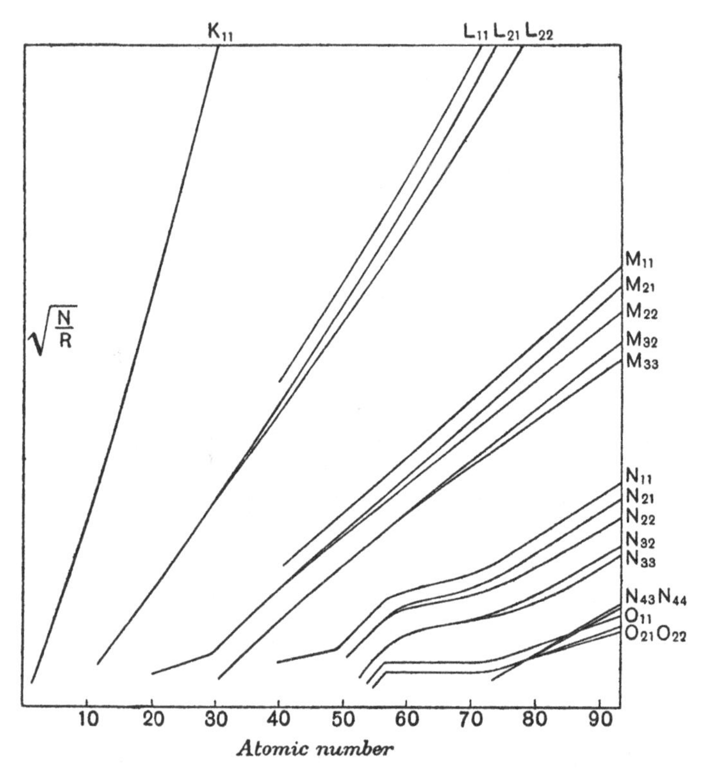

number of 32 electrons. In the Bohr-Coster diagram this again is indicated by curved lines up to $Z = 48$ and between $Z = 56$ and 70, but straight lines between $Z = 48$ and 56 and beyond $Z = 70$; the points of change appear as sudden bends in the curves.

31. *Multiple structure of the X-ray terms.*

In the Bohr-Coster diagram there is one K term for any element, 2 to 3 L terms, 3 to 5 M terms and so on. This multiplicity is exactly analogous to that of the *optical doublets*, both as to the number of terms and their mode of combination; hence the principal quantum numbers

$n = 1, 2, 3, \ldots$ are assigned to the $K, L, M, \ldots$ terms and quantum numbers l, j are used for the multiple structure of the terms. For the optical doublets the scheme is

$l \backslash j$	$\frac{1}{2}$	$\frac{3}{2}$	$\frac{5}{2}$
0	s		
1	p_1	p_2	
2		d_1	d_2

where $\Delta l = \pm 1$, $\Delta j = \pm 1$ or 0, and $l = 0, 1, 2, \ldots (n - 1)$.

Hence for the X-ray terms the schemes are:

$l \backslash j$	$\frac{1}{2}$
0	K_{11}

$(n = 1)$

$l \backslash j$	$\frac{1}{2}$	$\frac{3}{2}$
0	L_{11}	
1	L_{21}	L_{22}

$(n = 2)$

$l \backslash j$	$\frac{1}{2}$	$\frac{3}{2}$	$\frac{5}{2}$
0	M_{11}		
1	M_{21}	M_{22}	
2		M_{32}	M_{33}

$(n = 3)$

$l \backslash j$	$\frac{1}{2}$	$\frac{3}{2}$	$\frac{5}{2}$	$\frac{7}{2}$
0	N_{11}			
1	N_{21}	N_{22}		
2		N_{32}	N_{33}	
3			N_{43}	N_{44}

$(n = 4)$

and so on, with the same rules of combination: $\Delta l = \pm 1$, $\Delta j = \pm 1$ or 0.

These terms, 1 K term, 3 L terms, 5 M terms, 7 N terms, agree with the Bohr-Coster diagram. It may be noted that in this diagram the curves L_{11}, L_{21} are parallel, as are also the pairs M_{11}, M_{21}; M_{22}, M_{32}; N_{11}, N_{21}; N_{22}, N_{32}; N_{33}, N_{43}; the tables show that the members of a parallel pair have the same j but different l's.

32. *X-ray screening doublets.*

Since the curves L_{11}, L_{21} are parallel in the Bohr-Coster diagram, the difference of the $\sqrt{\frac{N}{R}}$ for the L_{11} and L_{21} terms is the same for all the elements; such a pair of terms gives rise to a 'screening' doublet (Hertz)[1]. They have the same j but a different l, so that they correspond to rosette orbits of different shapes (for the shape is determined by l); hence the screening effect of the core electrons is different for the two terms.

Thus for L_{11}, $$\frac{N}{R} = \frac{(Z - \sigma_2)^2}{2^2},$$

and for L_{21}, $$\frac{N}{R} = \frac{(Z - \sigma_2')^2}{2^2}.$$

Therefore the difference of these two $\sqrt{\frac{N}{R}}$'s is $\frac{\sigma_2 - \sigma_2'}{2}$, and is independent of Z; thus the parallelism of the curves in the Bohr-Coster diagram is accounted for. So for the two screening pairs of the M terms and the three screening pairs of the N terms; and so on.

33. *X-ray spin doublets.*

These are pairs of terms having the same l but different j's, for example L_{21}, L_{22}; M_{21}, M_{22}; M_{32}, M_{33}; N_{21}, N_{22}; N_{32}, N_{33}; N_{43}, N_{44} are pairs of terms each giving rise to a 'spin' doublet. The differences are due to the different orientations of the spin axis of the electron corresponding

[1] G. Hertz, Zs. f. Phys. 3, p. 19, 1920.

to $j = l \pm \frac{1}{2}$ (§ 23). Since for a given pair of terms the orbits have the same l they have the same rosette form and thus the same screening constant σ. Later theory will be given to show that the spin doublet separation is proportional to $(Z - \sigma)^4$, (§ 35); the screening doublet separation is, however, independent of Z. This is shown qualitatively in the Bohr-Coster diagram, where for large values of Z the spin separation $L_{21} \sim L_{22}$ is seen to be large compared with the screening separation $L_{11} \sim L_{21}$. The spin doublet does not become measurable until about $Z = 30$, when the two L levels split into three.

For the L terms we have the graphical representation given by Uhlenbeck and Goudsmit:

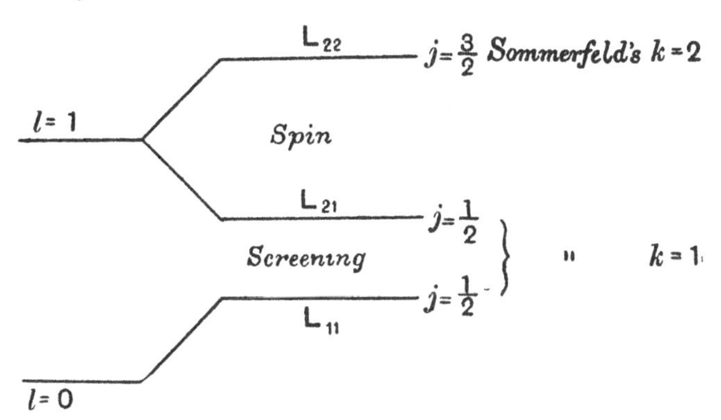

The original theory of Sommerfeld explained the spin doublet $L_{22} \sim L_{21}$ as a relativity effect where his quantum number k of the theory of the hydrogen fine structure[1] changed from 2 to 1 ($n = 2$ for the L lines); and then he could not really satisfactorily explain the screening doublet $L_{11} \sim L_{21}$ as that needed difference of orbital form for the L_{11} and L_{21} terms just when his k could not change. On

[1] Q.T.A. chap. VIII.

the theory of Uhlenbeck and Goudsmit, the old Sommerfeld relativity doublet of the X-ray spectra is due to the spin of the electron, and is the spin doublet of the new theory; variation of orbital form, such as relativity introduces, then accounts for the screening doublets.

34. *Optical doublets.*

In the case of the optical spectra of the alkalis the screening doublets vanish, as the screening number σ is the same for all orbits with different l's and the $\sigma_2 - \sigma_2'$ of § 32 is zero. This is because only for a very small part of the time in the orbital path does the series electron dive into the core[1]; while for the X-ray doublet the exciting electron is within the core all the time.

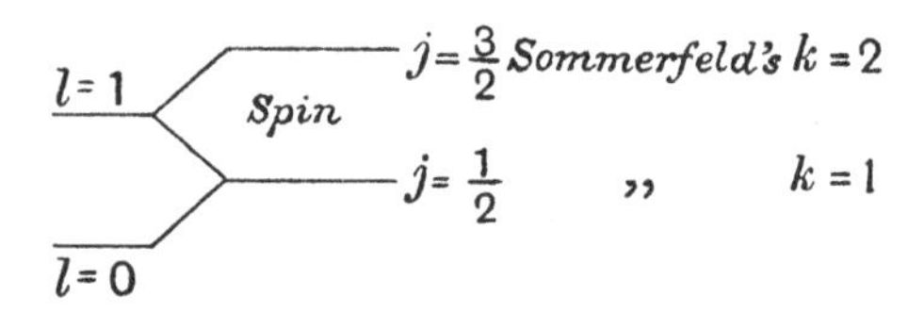

The optical doublets are entirely due to electron spin, and for a given l are determined by two j's equal to $l \pm \frac{1}{2}$.

The X-ray transitions.

The structure of the X-ray multiplets is determined by the conditions $\Delta l = \pm 1$, $\Delta j = \pm 1$ or 0 (§ 4), which must be satisfied for any transition between two levels.

The figure shows that the K_α and K_β lines are doublets and that the L_α line has seven components. By this scheme the nature of the multiple structure of any X-ray line can be found.

[1] Q.T.A. §§ 82, 91.

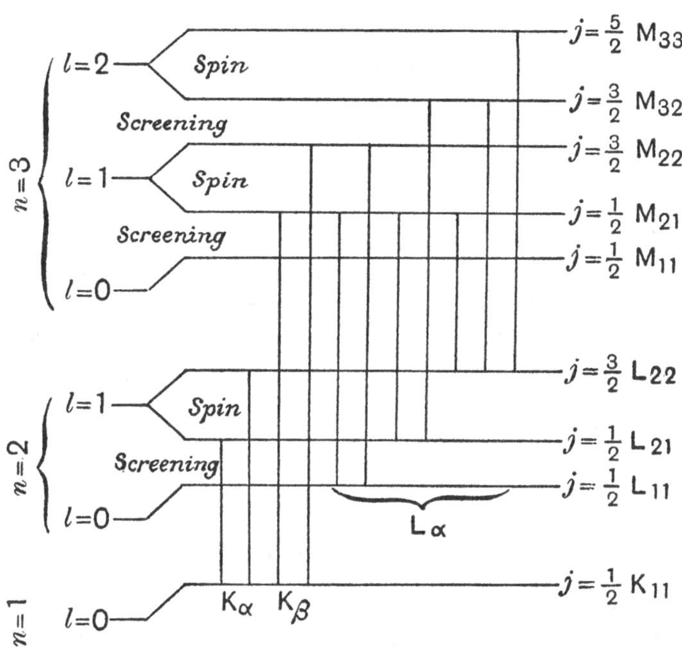

35. *The magnitude of the spin doublets as given by the new mechanics.*

Heisenberg and Jordan[1] using the new mechanics worked out the motion of a spinning electron ($-e$) moving under a Coulomb field due to a nucleus (Ze) taking into account the relativity mass variation with velocity. (The calculations are given later in § 127.)

The energy of an orbital state or multiplet term whose quantum numbers are n, l, j was found to be

$$\frac{C}{n^2}.Z^2 + \frac{C\alpha^2 Z^4}{n^3}\left\{\frac{j(j+1) - l(l+1) - s(s+1)}{2l(l+\frac{1}{2})(l+1)}\right\}$$
$$+ \frac{C\alpha^2 Z^4}{n^3}\left\{\frac{3}{4n} - \frac{1}{l+\frac{1}{2}}\right\} \quad \ldots\ldots(1),$$

[1] W. Heisenberg and P. Jordan, Zs. f. Phys. **37**, p. 263, **1926**.

where $C = Rch$, R being the Rydberg constant

$$\frac{2\pi^2 me^4}{ch^3} \equiv 109677$$

and α the fine structure constant $= \frac{2\pi e^2}{hc}$ (§ 21).

$$(\alpha^2 = 5{\cdot}31 \times 10^{-5}.)$$

The first term of (1) is the usual Balmer term, the second is the spin term and the third the relativity fine structure term. *The spin doublet separation* corresponds to different j's with the same l and is equal to the difference in the values of (1) when $j = l \pm \frac{1}{2}$ (cf. figure of § 33) and is therefore

$$\frac{C\alpha^2 Z^4}{n^3}\left\{\frac{(l+\frac{1}{2})(l+\frac{3}{2})-(l-\frac{1}{2})(l+\frac{1}{2})}{2l\,(l+\frac{1}{2})\,(l+1)}\right\} = \frac{C\alpha^2 Z^4}{n^3}\,\frac{1}{l\,(l+1)};$$

it has exactly the form of the Sommerfeld formula for the separation in the relativity fine structure for hydrogen with his k replaced by $l + 1$.

[The Sommerfeld term for the fine structure of hydrogen is $\frac{C\alpha^2}{n^3}\left(\frac{1}{k} - \frac{3}{4n}\right)$, so that the difference for a change $k \rightarrow k - 1$ is $\frac{C\alpha^2}{n^3}\,\frac{1}{k\,(k-1)}$.][1]

The Z of this theory is the effective Z after screening of the nucleus is allowed for and is thus the $Z - \sigma$ of § 33, so that the spin separation varies as $(Z - \sigma)^4$ in the notation of that paragraph.

The *screening doublet separation* corresponds to different l's with the same j, and is equal to the difference in the values of (1) when $l = j \pm \frac{1}{2}$.

When $l = j + \frac{1}{2}$, the expression (1), omitting the Balmer

[1] Q.T.A. § 60.

term which does not change with l, becomes, writing $s = \frac{1}{2}$,

$$\frac{C\alpha^2 Z^4}{n^3}\left[\frac{j(j+1)-(j+\frac{1}{2})(j+\frac{3}{2})-\frac{3}{4}}{2(j+\frac{1}{2})(j+1)(j+\frac{3}{2})}+\frac{3}{4n}-\frac{1}{j+1}\right]$$

$$=\frac{C\alpha^2 Z^4}{n^3}\left[-\frac{1}{2(j+\frac{1}{2})(j+1)}+\frac{3}{4n}-\frac{1}{j+1}\right]$$

$$=\frac{C\alpha^2 Z^4}{n^3}\left[-\frac{1}{j+\frac{1}{2}}+\frac{3}{4n}\right] \quad \text{.........................(2).}$$

When $l = j - \frac{1}{2}$, the expression becomes

$$\frac{C\alpha^2 Z^4}{n^3}\left[\frac{1}{2j(j+\frac{1}{2})}+\frac{3}{4n}-\frac{1}{j}\right]$$

$$=\frac{C\alpha^2 Z^4}{n^3}\left[-\frac{1}{j+\frac{1}{2}}+\frac{3}{4n}\right] \quad \text{............(3).}$$

(2) and (3) are the same, so that the screening doublet separation vanishes (cf. the first figure of § 34).

36. For the L terms of the X-ray spectra of the higher elements, the Z of § 35 (which is the $Z-\sigma$ of § 33) is different if l changes from $j+\frac{1}{2}$ to $j-\frac{1}{2}$, as the orbits are inside the core with shells of electrons outside them so that the screening number σ is greatly affected by the change of shape of the orbit, implied in the change of l.

The screening doublet is then dependent mainly on the Balmer term and is equal to $\frac{C}{n^2}(Z'^2 - Z^2)$, where Z', Z are the effective Z's corresponding to

$$l = j + \frac{1}{2}, \quad l = j - \frac{1}{2}$$

(cf. figure of § 33).

CHAPTER VII

THE PASCHEN-BACK EFFECT AND THE SOMMERFELD FORMULA; LANDÉ'S FORMULA FOR THE MAGNITUDE OF THE ALKALI DOUBLETS; THE TRIPLETS OF THE ALKALINE EARTHS

37. *The Paschen-Back effect.*

Every single line of a multiplet exhibits in a *weak* magnetic field a typical Zeeman effect whose separations are proportional to the field and are symmetrical with respect to the original line. (The field is regarded as weak when the Zeeman separations are of smaller order than the width of the original multiplet.) If the field is increased, the Zeeman lines of the different components of the multiplet spread out and overlap. Before this overlapping begins, an unsymmetrical distortion of the single Zeeman types shows itself, and this applies to the intensities as well as to the lie of the lines when the field becomes *strong*; certain lines flow together while others, losing their intensity, vanish; and finally the lines form a *normal* Zeeman triplet whose separation is large compared with the breadth of the original multiplet. This normalising of the Zeeman effect by strong fields was discovered by Paschen and Back[1].

38. *General theory of the Paschen-Back effect.*

The model of a multiplet state has coupled momenta S, L whose resultant is J, precessing round J. If the magnetic field is *weak* and so does not affect the internal coupling between S and L, its effect is to cause a precession of J about the axis of the field. This, as has

[1] F. Paschen and E. Back, Ann. der Phys. **39**, p. 897, **1912**.

been seen, gives the Zeeman terms mgo, where m is $\pm j$, $\pm (j-1)$, etc. But in *strong* fields, the coupling between L and S is broken down; each precesses independently round the axis of the field and has a quantised component along the field whose quantum numbers are respectively m_l, m_s. These give separate energy terms $m_l o$, $2m_s o$, on account of the double ratio of magnetic to mechanical moment for the spinning electron, and the Zeeman terms are $(m_l + 2m_s)\,o$.

Since l is an integer and $m_l = \pm l, \pm (l-1), \ldots,$ m_l is also an integer; since $s = \frac{1}{2}$, m_s can only be $\pm \frac{1}{2}$. Thus $m_l + 2m_s$ is always an integer and the Zeeman effect is *normal*, in accord with the observations of Paschen and Back.

39. *Sommerfeld's formula for the anomalous Zeeman effect of the sodium D doublet.*

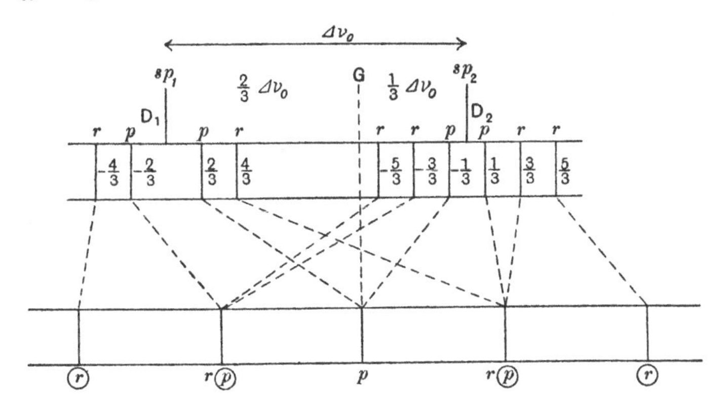

The figure shows the Paschen-Back transformation of the sodium D doublet, the upper figure giving the lines in a weak field and the lower one the lines in a strong field. The line D_2 is twice as intense as D_1 and G is the 'centre of gravity' of the intensity of the two lines. The terms p, r

enclosed with a circle lose their intensity so that the result of the strong field is the normal Zeeman triplet.

Sommerfeld[1] gave the quantum theory modification of Voigt's original classical formula[2] of the anomalous effect for this doublet and found

$$\nu - \nu_G = om - \frac{\omega}{2\,(2l+1)} \pm \sqrt{o^2 + \frac{4mo\omega}{2l+1} + \omega^2},$$

the upper sign holding for sp_1 and the lower for sp_2; also $\omega = \Delta\nu_0$, the original width of the doublet, and $o = eH/4\pi m_0 c$.

By considering the cases of a strong field $o \gg \omega$ and a weak field $o \ll \omega$, and expanding the square root approximately in the two cases, he was able to trace the changes shown in the diagram on p. 55 for the D doublet.

The theory has recently been given on the new mechanics by Heisenberg and Jordan, who agree with Sommerfeld's result[3]; the calculations are given later in § 124.

40. *Landé's formula for the magnitude of the alkali doublets.*

The doublet separation of the first line of the principal series of the alkalis increases with atomic number and becomes pronounced for the heavier metals of the series, increasing from ·25 Å for Li to 422·34 Å for Cs.

For an X-ray spin doublet corresponding to an effective nuclear charge $Z_i e$ and effective quantum number n_i, the separation is

$$\Delta\nu_i = \frac{Rc\alpha^2 Z_i^4}{n_i^3\, l(l+1)}, \text{ (§ 35)}.$$

For an optical doublet, Landé supposes that the time t_i spent in describing the inner orbital loop[4] within the core is small compared with the time t_a spent in describing the

[1] A. Sommerfeld, Zs. f. Phys. **8**, p. 257, **1922**.

[2] W. Voigt, Ann. der Phys. **41**, p. 403, **1913**, and **42**, p. 210, **1913**.

[3] W. Heisenberg and P. Jordan, Zs. f. Phys. **37**, p. 270, equation 20, **1926**.

[4] Q.T.A. p. 121 and § 147.

outer orbital loop. The latter is approximately a Keplerian ellipse described under an effective nuclear charge $Z_a e$, where $Z_a = 1$ for a neutral atom A, $Z_a = 2$ for A_+, $Z_a = 3$ for A_{++} and so on. Effectively t_a is the time of revolution in the orbit supposed completely Keplerian.

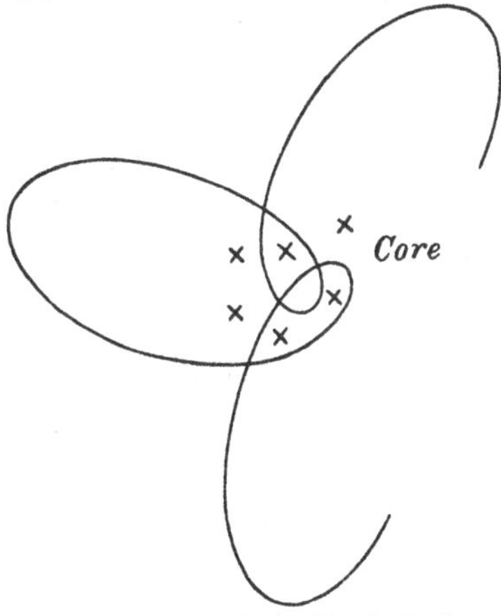

The inner orbital loop is described under the control of the nuclear charge modified just as in the case of the X-ray doublet.

Landé[1] supposes that $\Delta\nu$ for the optical lines is found approximately by multiplying the $\Delta\nu_i$ of the X-ray doublet by the fraction of the whole period that the electron spends in the inner loop (where it is under the action which gave rise to $\Delta\nu_i$), so that

$$\Delta\nu = \frac{t_i}{t_a}\Delta\nu_i .$$

Also the period t in the orbit[2] is $\frac{n^3 h^3}{4\pi^2 m e^4 Z^2}$, so that

$$\frac{t_i}{t_a} = \frac{n_i^3}{Z_i^2}\frac{Z_a^2}{n_a^3},$$

[1] A. Landé, Zs. f. Phys. **25**, p. 46, **1924**.
[2] Q.T.A. § 30.

where n_a is the 'effective' quantum number, being the $n - \alpha_k$ of the Rydberg formula[1].

Hence $\Delta\nu$ for the optical doublet $= \Delta\nu_i \dfrac{Z_a^2}{Z_i^2} \cdot \dfrac{n_i^3}{n_a^3}$

$$= \frac{R c \alpha^2 Z_i^4}{n_i^3 l(l+1)} \frac{Z_a^2}{Z_i^2} \frac{n_i^3}{n_a^3}$$

$$= \frac{R c \alpha^2 Z_i^2 Z_a^2}{n_a^3 l(l+1)}.$$

This is Landé's formula giving a quantitative result for the dependence of $\Delta\nu$ upon atomic number.

Bowen and Millikan[2] have tested this formula severely by experiments on Be_+, B_{++}, C_{+++}, N_{++++}, O_{+++++}; P_{++++}, S_{+++++}, Cl_{++++++}. These and the known results of Paschen[3] on Al_{++} and Fowler[4] on Si_{+++}, all of which have *one* series electron revolving about an inert gas core[5] as in the alkalis, have confirmed the Landé formula.

41. *Atoms with two series electrons.*

Such atoms are those of the *alkaline earths*, where two series electrons revolve about an inert gas core.

Each electron has its L (quantum number l) and S (quantum number s); these quantities for the two will be denoted by suffixes 1, 2.

The *normal* singulets and triplets of the alkaline earths can be accounted for by supposing that the coupling between L_1, L_2 and between S_1, S_2 is strong compared with that between L (the resultant of L_1 and L_2) and S (the resultant of S_1 and S_2). In that case L, S remain constant and precess about their resultant J.

Then since S is the resultant of S_1 and S_2,

$$|s_1 - s_2| \leqslant s \leqslant s_1 + s_2,$$

[1] Q.T.A. § 83.

[2] I. S. Bowen and R. A. Millikan, various papers in the Phys. Rev. vols. **22–26**, **1923–1925**.

[3] F. Paschen, Ann. der Phys. **71**, p. 142, **1923**.

[4] A. Fowler, Phil. Tr. Roy. Soc. A. **225**, p. 1, **1925**.

[5] Q.T.A. pp. 156, 157, § 92.

and since $s_1 = \frac{1}{2}$, $s_2 = \frac{1}{2}$, this condition is $0 \leqslant s \leqslant 1$. Hence s must be 0 or 1.

So since J is the resultant of L and S,

$$| l - s | \leqslant j \leqslant l + s.$$

Therefore if $s = 0$, j must $= l$ and we have the *singulet* terms, and if $s = 1$,

$$| l - 1 | \leqslant j \leqslant l + 1,$$

so that $j = l - 1$, l, or $l + 1$ which are the *triplet* terms. This is all in agreement with the singulet and triplet tables of § 5.

42. The above is only a special case of the theory of coupling[1] of the vectors concerned, but it is beyond the purpose of this book to pursue these questions further. A full account of the 'confused' (*verschobenen*) terms of the calcium spectrum (due to excited states of the series electrons), and of spectra of higher multiplicity has been given by Hund[2] in his recent book on line spectra.

The whole question too is affected by the new 'resonance' theory of Heisenberg which solves the problem of the helium spectrum, and which will be considered in detail in chapter XXVI.

[1] G. E. Uhlenbeck and S. Goudsmit, Zs. f. Phys. **35**, p. 618, **1926**.
[2] 'Linienspektren,' by F. Hund, Berlin, **1927**.

CHAPTER VIII

THE NEW QUANTUM KINEMATICS OF HEISENBERG; MATRICES AND NON-COMMUTATIVE MULTIPLICATION

43. *The new quantum kinematics of Heisenberg*[1].

On the classical theory a *real* coordinate q is represented by a Fourier series, i.e.

$$q(t) = a_0 + a_1 \cos 2\pi\nu t + a_2 \cos 2(2\pi\nu t) + a_3 \cos 3(2\pi\nu t) + \dots$$
$$+ b_1 \sin 2\pi\nu t + b_2 \sin 2(2\pi\nu t) + b_3 \sin 3(2\pi\nu t) + \dots,$$

where the a's and b's are real constants. The series represents a set of waves of frequencies ν, 2ν, 3ν, ...

Writing exponential forms for the sine and cosine,

$$q(t) = a_0 + \frac{1}{2}(a_1 - ib_1)\, e^{2\pi i\nu t} + \frac{1}{2}(a_2 - ib_2)\, e^{2(2\pi i\nu t)} + \dots$$
$$+ \frac{1}{2}(a_1 + ib_1)\, e^{-2\pi i\nu t} + \frac{1}{2}(a_2 + ib_2)\, e^{-2(2\pi i\nu t)} + \dots,$$

or $$q(t) = q_0 + q_1 e^{2\pi i\nu t} + q_2 e^{2(2\pi i\nu t)} + \dots$$
$$+ q_{-1} e^{-2\pi i\nu t} + q_{-2} e^{-2(2\pi i\nu t)} + \dots, \text{ suppose,} \quad \dots\dots(1),$$

where $q_{-\tau}$ is the conjugate of q_τ; or $q_{-\tau} = q_\tau^*$, where q_τ^* denotes the conjugate of q_τ.

Hence $$q(t) = \sum_{-\infty}^{\infty}{}_{\tau}\, q_\tau e^{\tau(2\pi i\nu t)} \quad \dots\dots\dots\dots\dots\dots(2).$$

On the quantum theory Heisenberg generalises this so as to express $q(t)$ in terms of the amplitudes and frequencies of the *spectral lines* related to $q(t)$.

Such frequencies are known not to be of the classical

[1] W. Heisenberg, Zs. f. Phys. **33**, p. 879, July 1925.

type $\tau\nu$, but to be expressible (by the Ritz combination principle)[1] as the differences of the 'term' series

$$T_1, T_2, T_3, \ldots.$$

We write $\nu(nm)$ as the frequency of the spectral line corresponding to the terms T_n and T_m, so that

$$\nu(nm) = T_n - T_m \quad \ldots\ldots\ldots\ldots\ldots\ldots(3).$$

Thus $\nu(nm) = -\nu(mn)$, and $\nu(nn) = 0$.

The aggregate of all the possible waves is

$$\begin{pmatrix} q(11)\, e^{2\pi i\nu(11)t}, & q(12)\, e^{2\pi i\nu(12)t}, & q(13)\, e^{2\pi i\nu(13)t}, & \ldots \\ q(21)\, e^{2\pi i\nu(21)t}, & q(22)\, e^{2\pi i\nu(22)t}, & q(23)\, e^{2\pi i\nu(23)t}, & \ldots \\ q(31)\, e^{2\pi i\nu(31)t}, & q(32)\, e^{2\pi i\nu(32)t}, & q(33)\, e^{2\pi i\nu(33)t}, & \ldots \\ \ldots\ldots, & \ldots\ldots, & \ldots\ldots, & \ldots \end{pmatrix} \ldots(4).$$

These terms form a '*table*,' which is denoted by

$$(q(nm)\, e^{2\pi i\nu(nm)t})$$

and the table represents the variable q, so that

$$q = (q(nm)\, e^{2\pi i\nu(nm)t}),$$

and $q(nm)$ is the amplitude corresponding to the frequency $\nu(nm)$.

The diagonal terms of the table (4), whose amplitudes are $q(11)$, $q(22)$, ..., are constants, since $\nu(11)$, $\nu(22)$, ... are zero; these terms are a set corresponding to the term q_0 of the classical series (1).

Just as for the Fourier series (1) to represent a real q, we must have $q_{-\tau} = q_\tau^*$, so here must $q(mn) = q^*(nm)$, so that $q(nm)\, q(mn) = |q(nm)|^2$.

44. *Use of matrices.*

In building up the laws of calculation for quantum magnitudes q expressed by tables of this kind, Heisenberg used the analogies of classical mechanics as far as possible; for the theory of conditionally periodic systems of classical mechanics has accounted for so much of quantum phenomena that he sought to retain the old mechanical laws and supplement them by a new scheme of kinematics.

[1] Q.T.A. § 15.

He considered the simplest case, that of Planck's oscillator, for which the potential energy is $\frac{1}{2}kq^2$; how was the table for q^2 to be found when that for q was known?

On the classical theory

$$q = \sum_{-\infty}^{\infty}{}_{\tau}\, q_\tau e^{2\pi i\tau\nu t}, \text{ or } \Sigma_\tau q_\tau \exp 2\pi i\tau\nu t$$

and $$q^2 = (\Sigma_\sigma q_\sigma \exp 2\pi i\sigma\nu t)(\Sigma_{\sigma'} q_{\sigma'} \exp 2\pi i\sigma'\nu t)$$
$$= \Sigma_\sigma \Sigma_{\sigma'} q_\sigma q_{\sigma'} \exp 2\pi i(\sigma + \sigma')\nu t;$$

and writing $$\sigma + \sigma' = \tau, \text{ or } \sigma' = \tau - \sigma,$$
$$q^2 = \Sigma_\tau \Sigma_\sigma q_\sigma q_{\tau-\sigma} \exp 2\pi i\tau\nu t$$
$$= \Sigma_\tau Q_\tau \exp 2\pi i\tau\nu t,$$

where $$Q_\tau = \Sigma_\sigma q_\sigma q_{\tau-\sigma}.$$

Thus the series of magnitudes $Q_\tau \exp 2\pi i\tau\nu t$ represents q^2 in the same way that the series of magnitudes $q_\tau \exp 2\pi i\tau\nu t$ represents q; the Fourier series for q^2 has the same frequencies as that for q, and $Q_\tau = \sum_{-\infty}^{\infty}{}_\sigma\, q_\sigma q_{\tau-\sigma}$.

Heisenberg put himself the question—can a multiplication rule for the $q(nm)$'s be found by which from a table for q a new table for q^2 can be derived without new frequencies appearing, as in the classical theory? The answer was that the table is to be regarded as a *matrix* in the theory of algebra.

In algebra, the product of two matrices $a \equiv (a(nm))$

$$\equiv \begin{pmatrix} a(11) & a(12) & a(13) \ldots \\ a(21) & a(22) & a(23) \ldots \\ a(31) & a(32) & a(33) \ldots \\ \ldots\ldots & \ldots\ldots & \ldots\ldots \end{pmatrix}, \text{ and } b \equiv (b(nm)),$$

is a matrix $c \equiv (c(nm))$, where

$$c(nm) = \Sigma_k a(nk)\, b(km).$$

Thus for two quantum tables,

$$x \equiv (x\,(nm) \exp 2\pi i\nu\,(nm)\,t)$$
$$y \equiv (y\,(nm) \exp 2\pi i\nu\,(nm)\,t),$$
$$xy = (\Sigma_k\, x\,(nk) \exp 2\pi i\nu\,(nk)\,t\,.\,y\,(km) \exp 2\pi i\nu\,(km)\,t)$$
$$= (\Sigma_k\, x\,(nk)\,y\,(km) \exp 2\pi i\,\{\nu\,(nk) + \nu\,(km)\}\,t).$$

But $\nu\,(nk) + \nu\,(km) = (T_n - T_k) + (T_k - T_m)$,

from equation (3),

$$= T_n - T_m$$
$$= \nu\,(nm),$$

so that $xy = (\Sigma_k\, x\,(nk)\,y\,(km) \exp 2\pi i\nu\,(nm)\,t)$,

and the new table has the same frequencies as the tables for x and y.

Thus matrix representation of the quantum variables depends upon the Ritz combination principle; and if $z = xy$,

$$z\,(nm) = \Sigma_k\, x\,(nk)\,y\,(km).$$

The exponential time factors are usually not written in the matrices, but it is understood all the while that a term written just as $z\,(nm)$ means $z\,(nm) \exp 2\pi i\nu\,(nm)\,t$.

As the matrices represent real quantities,

$$x\,(nm) = x^*\,(mn),$$

i.e. the matrices are of the Hermite type.

45. *Some properties of matrices.*

(i) $a = b$ means $a\,(nm) = b\,(nm)$,

$a = b + c$ means $a\,(nm) = b\,(nm) + c\,(nm)$,

$a = bc$ means $a\,(nm) = \Sigma_k\, b\,(nk)\,c\,(km)$.

The laws $(ab)\,c = a\,(bc)$ and $a\,(b + c) = ab + ac$ hold, but *ab is not in general equal to ba*; in cases where $ab = ba$, a and b are said to be 'commutable.' Hence the *commutative law of multiplication does not hold* in general for quantum variables.

(ii) The *unit matrix*

$$1 \equiv \begin{pmatrix} 1 & 0 & 0 & 0 \ldots \\ 0 & 1 & 0 & 0 \ldots \\ 0 & 0 & 1 & 0 \ldots \\ \ldots & \ldots & \ldots & \ldots \ldots \end{pmatrix} \text{ or } (\delta(nm)),$$

where
$$\delta(nm) = 1 \text{ for } n = m$$
$$= 0 \text{ for } n \neq m.$$

Also $a1 = (\Sigma_k \, a(nk) \, \delta(km))$

$= (a(nm))$, since $\delta(km) = 0$ for every k except m,

$= a.$

$\therefore \; a1 = a.$

So $1a = a$, and 1 commutes with any matrix a.

The reciprocal matrix a^{-1} is defined by $a^{-1}a = 1$.

Since $a^{-1}a = 1$, then $a(a^{-1}a) = a1 = 1a$.

$$\therefore \; (aa^{-1})\, a = 1a.$$
$$\therefore \; aa^{-1} = 1.$$

Thus $a^{-1}a = 1 = aa^{-1}$, so that a^{-1} and a commute.

(iii) A *diagonal matrix* is one of the form

$$\begin{pmatrix} a(11) & 0 & 0 & \ldots \\ 0 & a(22) & 0 & \ldots \\ 0 & 0 & a(33) & \ldots \\ \ldots & \ldots & \ldots & \ldots \end{pmatrix} \text{ or } (a(nm)\, \delta(nm)).$$

As in the Heisenberg matrices there is a time factor $e^{2\pi i\nu(nm)t}$ multiplying each $q(nm)$, the diagonal terms of any matrix are independent of t, so that a diagonal matrix corresponds to a constant of the classical theory. But the converse is not necessarily true—a constant need not always be a diagonal matrix. If certain of the frequencies are zero, say for example $\nu(23) = 0$ and $\nu(14) = 0$, the terms (23), (32), (14), (41) are constants, so that a constant might be represented by a matrix containing these non-diagonal terms as well as the diagonal terms;

and so not be a diagonal matrix. This case of vanishing frequencies is the well-known case of a *degenerate*[1] system of the older quantum theory, which arises for instance in the case of the Keplerian orbit. For non-degenerate systems, a constant must be a diagonal matrix.

46. *The quantum condition for a single variable.*

Let q be the coordinate and p the conjugate momentum. On the classical theory

$$J = \oint p\,dq = \int_0^{\frac{1}{\nu}} p\dot{q}\,dt,$$

since $\frac{1}{\nu}$ is a period.

If
$$p = \Sigma p_\tau \exp 2\pi i \tau \nu t,$$
and
$$q = \Sigma q_\tau \exp 2\pi i \tau \nu t,$$
then
$$p\dot{q} = \Sigma_\tau p_\tau \exp 2\pi i \tau \nu t \; \Sigma_\tau 2\pi i \nu \tau \, q_\tau \exp 2\pi i \tau \nu t$$
$$= \Sigma_\tau Q_\tau \exp 2\pi i \tau \nu t,$$
where
$$Q_\tau = \Sigma_k p_k \, 2\pi i \nu \, (\tau - k) \, q_{\tau - k}, \text{ as in § 44.}$$

$$\therefore \int_0^{\frac{1}{\nu}} p\dot{q}\,dt = \int_0^{\frac{1}{\nu}} \Sigma_\tau Q_\tau \exp 2\pi i \tau \nu t \, dt$$
$$= \frac{Q_0}{\nu},$$

since only the constant term of the series yields anything.

But
$$Q_0 = \Sigma_k p_k \, (2\pi i \nu) \, (-k) \, q_{-k},$$
$$\therefore \; J = -2\pi i \, \Sigma_k k p_k q_{-k}$$
$$= -2\pi i \, \Sigma_\tau \tau p_\tau q_{-\tau}, \text{ writing } \tau \text{ for } k \ldots (1).$$

But[2] $\frac{\partial F}{\partial J}$ of the classical mechanics translates by the correspondence principle into $\frac{\Delta F}{\Delta J}$ or $\frac{\Delta F}{\tau h}$ of the older

[1] Q.T.A. §§ 55, 56.

[2] Q.T.A. §§ 57, 151.

quantum theory, where ΔF is the change of a function F due to J changing by τh.

A classical amplitude $A_\tau(n)$ corresponds to a quantum theory amplitude $A(n, n-\tau)$.

The equation (1) is therefore differentiated with respect to J to bring it to a form suitable for the use of the correspondence principle, and we have

$$1 = -2\pi i \Sigma_\tau \tau \frac{\partial}{\partial J}(p_\tau q_{-\tau}) \ldots\ldots\ldots\ldots\ldots (2).$$

Now $\frac{\partial}{\partial J}\{A_\tau(n)\}$ translates into $\frac{\Delta}{\tau h}\{A(n, n-\tau)\}$ or

$$\frac{A(n+\tau, n+\tau-\tau) - A(n, n-\tau)}{\tau h},$$

so that $\tau\frac{\partial}{\partial J}\{A_\tau(n)\}$ of the classical theory goes into

$$\frac{1}{h}\{A(n+\tau, n) - A(n, n-\tau)\}$$

of the quantum theory.

Thus $\frac{\partial}{\partial J}\{p_\tau(n)\, q_{-\tau}(n)\}$ of equation (2) translates into

$$\frac{\Delta}{\tau h}\{p(n, n-\tau)\, q(n-\tau, n)\},$$

so that $\tau\frac{\partial}{\partial J}\{p_\tau(n)\, q_{-\tau}(n)\}$ becomes

$$\frac{1}{h}[p(n+\tau, n+\tau-\tau)\, q(n+\tau-\tau, n+\tau) \\ - p(n, n-\tau)\, q(n-\tau, n)],$$

or $\frac{1}{h}[p(n+\tau, n)\, q(n, n+\tau) - p(n, n-\tau)\, q(n-\tau, n)]$.

Hence substituting in equation (2) we have

$$\frac{ih}{2\pi} = \Sigma_\tau\{q(n, n+\tau)\, p(n+\tau, n) - p(n, n-\tau)\, q(n-\tau, n)\}$$

$$= \Sigma_k\{q(nk)\, p(kn) - p(nk)\, q(kn)\}, \text{ since } n \text{ is large.}$$

But the right-hand side is the (nn) term of the matrix $qp - pq$ or the nth term of the diagonal. Thus all the diagonal terms of the matrix $qp - pq$ are equal to $\frac{ih}{2\pi}$, where h is Planck's constant.

This clue was generalised into

$$qp - pq = \frac{ih}{2\pi}1 \quad\text{..................(3)},$$

where 1 is the unit matrix, the step to be justified by results.

Thus the quantum condition of the older theory leads to an algebraic equation which makes it possible to perform calculations in the new and wider scheme of dynamics where the classical equations are used, but with a non-commutative multiplication law for the variables. The difference between pq and qp is known in terms of Planck's constant h. As $h \to 0$, $pq \to qp$ and we pass over to the classical mechanics.

47. *Matrices infinite.*

The Heisenberg matrices have an infinite number of terms.

This follows from the equation

$$\Sigma_k \{q(nk)\, p(kn) - p(nk)\, q(kn)\} = \frac{ih}{2\pi};$$

for if the number of states n were finite, then

$$\Sigma_n \Sigma_k \{q(nk)\, p(kn) - p(nk)\, q(kn)\} = 0,$$

since for a *finite* sum the order of the terms can be altered and they cancel in pairs.

But $\Sigma_n \frac{ih}{2\pi}$ would be $\frac{nih}{2\pi}$, and so would not be zero.

48. This stage of the theory was reached in September 1925 by Born and Jordan[1], who had developed Heisenberg's original speculations put forward in the preceding July.

Early in November, Paul Dirac[2], also starting from Heisenberg's new idea, had worked out a theory on somewhat different lines; he discovered that the quantum conditions for a multiply periodic system could be interpreted in terms of the well-known *Poisson bracket* expressions of the classical mechanics.

[1] M. BORN and P. JORDAN, Zs. f. Phys. **34**, p. 858, Sept. **1925**. This paper will be referred to as 'Quantenmechanik I' or 'Q.M. I.'

[2] P. A. M. DIRAC, Proc. Roy. Soc. A. **109**, p. 642, Nov. **1925**.

CHAPTER IX

THE THEORY OF DIRAC; USE OF POISSON BRACKETS; THE ENERGY LAW AND BOHR'S FREQUENCY CONDITION

49. *Properties of Poisson brackets.*

(i) In the classical dynamics, if x, y are functions of the coordinates $q_1 \dots q_s$ and the momenta $p_1 \dots p_s$, the expression

$$\sum_{1}^{s}{}_k \left(\frac{\partial x}{\partial q_k}\frac{\partial y}{\partial p_k} - \frac{\partial x}{\partial p_k}\frac{\partial y}{\partial q_k}\right)$$

is denoted by $[xy]$ and is called a *Poisson bracket.*

Directly from this definition it follows that

$$[p_r p_s] = 0, \quad [q_r q_s] = 0,$$

and $[q_r p_s] = \delta(rs)$, where $\delta(rs) = 1$, when $r = s$,
$= 0$, ,, $r \neq s$.

(ii) They have also the properties

$$[x + y, z] = [xz] + [yz],$$
$$[xy] = -[yx],$$
$$[xy, z] = [xz]\, y + x\,[yz].$$

By the use of these equations $[xy]$ can be built up from the expressions $[q_r p_s]$, $[p_r p_s]$, $[p_r q_s]$, if x, y can be expressed as power series in the p's and q's.

For instance

$$[q^2, p^2] = [q \cdot q, p^2] = [qp^2]\, q + q\,[qp^2],$$

and

$$[qp^2] = -[pp, q] = -p\,[pq] - [pq]\, p$$
$$= 2p, \text{ since } [qp] = 1 \text{ by (i).}$$

$$\therefore \quad [q^2, p^2] = 2pq + 2qp.$$

(iii) *Canonical transformation*[1].

If we change the variables from p, q to new variables P, Q by a canonical transformation (which leaves the

[1] 'Analytical Dynamics,' by E. T. Whittaker, Cambridge, **1917**.

Hamiltonian equations unchanged in form) it is a known result of the classical mechanics that

$$\Sigma_k \left(\frac{\partial x}{\partial q_k}\frac{\partial y}{\partial p_k} - \frac{\partial x}{\partial p_k}\frac{\partial y}{\partial q_k}\right) = \Sigma_k \left(\frac{\partial x}{\partial Q_k}\frac{\partial y}{\partial P_k} - \frac{\partial x}{\partial P_k}\frac{\partial y}{\partial Q_k}\right),$$

or that $[xy]$ is the same for the new as for the old variables.

Hence $[P_r P_s]$ with respect to $p, q = [P_r P_s]$ with respect to $P, Q = 0$, by § 49, (i), and so forth.

Hence $[P_r P_s] = 0$, $[Q_r Q_s] = 0$, $[Q_r P_s] = \delta\,(rs)$, with respect to the p's and q's, are the conditions for P, Q to be canonical variables.

50. *Dirac's procedure*[1].

Dirac considers the expression $xy - yx$ of the quantum theory and finds the corresponding formula of the classical theory to which it must tend for large quantum numbers; in the latter theory x, y are functions of the coordinates $q_1 \dots q_s$ and momenta $p_1 \dots p_s$ of a multiply periodic system of s degrees of freedom.

Thus on the classical theory,

$$x = \Sigma_{\alpha_1 \dots \alpha_s}\, x\,(\alpha_1 \dots \alpha_s, J_1 \dots J_s) \exp 2\pi i\,(\alpha_1\nu_1 + \dots + \alpha_s\nu_s)\,t,$$

where $\alpha_1 \dots \alpha_s$ are integers, and the J's and the νt's are the action and angle variables[2]; or

$$x = \Sigma_\alpha\, x\,(\alpha, J) \exp 2\pi i\,(\alpha\nu)\,t, \text{ for brevity} \ldots\ldots (1).$$

The stationary states correspond to

$$J_1 = n_1 h, \quad J_2 = n_2 h, \; \dots .$$

On the quantum theory, x is the aggregate of terms

$$x\,(n, n - \alpha) \exp 2\pi i \nu\,(n, n - \alpha)\,t,$$

where $x\,(n, n - \alpha)$ is the amplitude corresponding to the frequency $\nu\,(n, n - \alpha)$ due to a transition from the state n (quantum numbers $n_1 \dots n_s$) to a state $(n - \alpha)$ (quantum numbers $n_1 - \alpha_1, n_2 - \alpha_2, \dots n_s - \alpha_s$).

[1] P. A. M. Dirac, Proc. Roy. Soc. A. **109**, p. 642, Nov. **1925**.
[2] Q.T.A. §§ 49 to 52.

Then for large quantum numbers n

$$x(n, n-\alpha) \rightarrow x(\alpha, J),$$

by the correspondence principle.

Now the nm component of the matrix $xy - yx$ is

$$\Sigma_k \{x(nk)\, y(km) - y(nk)\, x(km)\} \exp 2\pi i\nu(nm)\, t,$$

so that Dirac considered its equivalent for large n, m:

$$\underset{\alpha+\beta=n-m}{\Sigma_\alpha \, \Sigma_\beta} \{x(n, n-\alpha)\, y(n-\alpha, n-\alpha-\beta) - y(n, n-\beta) \times x(n-\beta, n-\beta-\alpha)\} \exp 2\pi i\nu(n, n-\alpha-\beta)\, t.$$

This expression is equal to

$$\Sigma_\alpha \Sigma_\beta \{[x(n, n-\alpha) - x(n-\beta, n-\beta-\alpha)]\, y(n-\alpha, n-\alpha-\beta) - [y(n, n-\beta) - y(n-\alpha, n-\alpha-\beta)]\, x(n-\beta, n-\beta-\alpha)\} \times \exp 2\pi i\nu(n, n-\alpha-\beta)\, t \;\ldots\ldots(2).$$

But $\dfrac{\Delta}{h} x(n, n-\alpha)$ translates into $\Sigma_r \tau_r \dfrac{\partial}{\partial J_r} x(\alpha, J)$, (cf. § 46), where Δ is the change due to each n_r becoming $n_r + \tau_r$; also the n's are large, so that

$$x(n, n-\alpha) = x(n+l, n+l-\alpha),$$

where l is any number not large.

Using these, the expression (2) translates into

$$\underset{\alpha+\beta=n-m}{\Sigma\Sigma} \sum_{r=1}^{r=s} h \left[\beta_r \left\{\frac{\partial}{\partial J_r} x(\alpha, J)\right\} y(\beta, J) - \alpha_r \left\{\frac{\partial}{\partial J_r} y(\beta, J)\right\} x(\alpha, J)\right] \exp 2\pi i \{(\alpha\nu)\, t + (\beta\nu)\, t\},$$

because $\nu(n, n-\alpha-\beta)$ translates into

$$(\alpha_1 + \beta_1)\,\nu_1 + \ldots + (\alpha_s + \beta_s)\,\nu_s$$

of the classical theory $= (\alpha\nu) + (\beta\nu)$.

Again the angle variable w_r canonical to the action variable J_r is $w_r = \nu_r t + \epsilon_r$, where ϵ_r is a phase constant.

$$\therefore \quad \frac{\partial}{\partial w_r} = \frac{\partial}{\partial(\nu_r t)}.$$

$$\therefore \quad \frac{\partial}{\partial w_r}\{y(\beta, J)\exp 2\pi i(\beta\nu)t\}$$

$$= \frac{\partial}{\partial(\nu_r t)}\{y(\beta, J)\exp 2\pi i(\beta_1\nu_1 + \ldots + \beta_r\nu_r + \ldots)t\}$$

$$= 2\pi i\beta_r\, y(\beta, J)\exp 2\pi i(\beta\nu)t.$$

Thus the nm component of the matrix $xy - yx$ translates into

$$\frac{h}{2\pi i}\sum_\alpha\sum_\beta{}_{\alpha+\beta=n-m}\;\Sigma_r\left[\frac{\partial}{\partial J_r}\{x(\alpha, J)\exp 2\pi i(\alpha\nu)t\}\right.$$

$$\times\frac{\partial}{\partial w_r}\{y(\beta, J)\exp 2\pi i(\beta\nu)t\} - \frac{\partial}{\partial J_r}\{y(\beta, J)\exp 2\pi i(\beta\nu)t\}$$

$$\left.\times\frac{\partial}{\partial w_r}\{x(\alpha, J)\exp 2\pi i(\alpha\nu)t\}\right],$$

so that the matrix itself becomes

$$\frac{h}{2\pi i}\Sigma_r\left(\frac{\partial x}{\partial J_r}\frac{\partial y}{\partial w_r} - \frac{\partial y}{\partial J_r}\frac{\partial x}{\partial w_r}\right), \text{ using (1),}$$

$$= \frac{ih}{2\pi}\Sigma_r\left(\frac{\partial x}{\partial w_r}\frac{\partial y}{\partial J_r} - \frac{\partial x}{\partial J_r}\frac{\partial y}{\partial w_r}\right)$$

$$= \frac{ih}{2\pi}\Sigma_r\left(\frac{\partial x}{\partial q_r}\frac{\partial y}{\partial p_r} - \frac{\partial x}{\partial p_r}\frac{\partial y}{\partial q_r}\right), \text{ by § 49 (iii)}$$

$$= \frac{ih}{2\pi}[xy].$$

Thus for large quantum numbers $xy - yx$ corresponds to $\frac{ih}{2\pi}[xy]$ of the classical theory, where $[xy]$ is the Poisson bracket of x and y. This result is taken over into the quantum theory and generalised thus: 'The difference between the Heisenberg products of two quantum variables is $\frac{ih}{2\pi}$ times their Poisson bracket expression, where the Poisson brackets are assumed to obey the same rules and to be calculated formally as in the classical theory.'

This is Dirac's new postulate for two quantum magnitudes x, y; it runs

$$xy - yx = \frac{ih}{2\pi}[xy] \quad\text{.................}(3).$$

He calls x and y q-numbers, as distinct from ordinary numbers, which he calls c-numbers.

51. *The quantum conditions for several variables.*

Dirac takes over into the quantum theory the equations

$$[q_r q_s] = 0,\ [p_r p_s] = 0,\ [q_r p_r] = 1,\ [q_r p_s] = 0,\ r \neq s.$$

Using (3) of the last article, these results are

$$\left.\begin{array}{l} \left.\begin{array}{l} q_r q_s - q_s q_r = 0 \\ p_r p_s - p_s p_r = 0 \end{array}\right\} \text{for all } r \text{ and } s \\ q_r p_s - p_s q_r = 0, \text{ if } r \neq s \\ q_r p_r - p_r q_r = \frac{ih}{2\pi} 1 \end{array}\right\} \quad\text{.........}(1).$$

These are the quantum conditions for a system with several variables and are the generalisation of the quantum condition of § 46. They were first given by Dirac[1] and were shortly afterwards found independently by Born, Heisenberg and Jordan[2].

52. *The general equation of motion.*

The methods of Heisenberg and Dirac are both used, and supplement one another.

On the classical theory, if x is a function of the p's and q's,

$$\dot{x} = \Sigma_k \left(\frac{\partial x}{\partial q_k}\dot{q}_k + \frac{\partial x}{\partial p_k}\dot{p}_k\right).$$

But if H is the Hamiltonian,

$$\frac{\partial H}{\partial p_k} = \dot{q}_k,\ \frac{\partial H}{\partial q_k} = -\dot{p}_k,$$

so that
$$\dot{x} = \Sigma_k \left(\frac{\partial x}{\partial q_k}\frac{\partial H}{\partial p_k} - \frac{\partial x}{\partial p_k}\frac{\partial H}{\partial q_k}\right).$$

$$\therefore\ \dot{x} = [xH].$$

[1] P. A. M. Dirac, Proc. Roy. Soc. A. **109**, p. 642, Nov. **1925**.

[2] M. Born, W. Heisenberg and P. Jordan, Zs. f. Phys. **35**, p. 557, Nov. **1925**. This will be referred to as 'Quantenmechanik II' or 'Q.M. II.'

This equation is in a form which can be taken directly over into the quantum theory as

$$\dot{x} = [xH] \quad \text{.....................(1).}$$

Using the Dirac equation $xy - yx = \frac{ih}{2\pi}[xy]$, equation (1) takes the Heisenberg form

$$xH - Hx = \frac{ih}{2\pi}\dot{x} \quad \text{..................(2).}$$

The interpretation of $\dot{x}$ now is that it is the matrix whose components are the time differential coefficients of the components of x, so that as x has constituents

$$x\,(nm) \exp 2\pi i\nu\,(nm)\,t,$$

$\dot{x}$ has constituents

$$2\pi i\nu\,(nm)\,x\,(nm) \exp 2\pi i\nu\,(nm)\,t,$$

or, omitting as usual the time factor, we write

$$\dot{x} = (2\pi i\nu\,(nm)\,x\,(nm)).$$

53. *The Hamiltonian equations.*

On the classical theory, the Poisson bracket $[p_k x]$, where x is a function of the p's and q's, is equal to

$$\Sigma_r \left(\frac{\partial p_k}{\partial q_r}\frac{\partial x}{\partial p_r} - \frac{\partial p_k}{\partial p_r}\frac{\partial x}{\partial q_r}\right).$$

$$\left.\begin{aligned} \therefore\quad [p_k x] &= -\frac{\partial p_k}{\partial p_k}\frac{\partial x}{\partial q_k} = -\frac{\partial x}{\partial q_k} \\ \text{So}\qquad [q_k x] &= \frac{\partial x}{\partial p_k} \end{aligned}\right\} \quad \text{...........(1).}$$

$$\text{Writing}\quad x = H,\quad \left.\begin{aligned} [p_k H] &= -\frac{\partial H}{\partial q_k} = \dot{p}_k \\ [q_k H] &= \frac{\partial H}{\partial p_k} = \dot{q}_k \end{aligned}\right\}.$$

These are the Hamiltonian equations and they are taken over unchanged into the quantum theory as

$$\left.\begin{aligned}\dot{p} &= [pH]\\ \dot{q} &= [qH]\end{aligned}\right\}, \text{ for each } p \text{ and } q.$$

They are special forms of the general equation $\dot{x} = [xH]$ of § 52, with $x = p$ and $x = q$.

The corresponding Heisenberg[1] equations are

$$\left.\begin{aligned}pH - Hp &= \frac{ih}{2\pi}[pH] = \frac{ih}{2\pi}\dot{p}\\ qH - Hq &= \frac{ih}{2\pi}\dot{q}\end{aligned}\right\}.$$

54. If x commutes with p and with q, it commutes with a function of p, q.

To show this, we deduce that if $[px] = 0$, and $[qx] = 0$ (for these mean, by the Dirac equation, that $px = xp$ and $qx = xq$), then $[f(p, q), x] = 0$.

But $[px] = 0$ means $\dfrac{\partial x}{\partial q} = 0$ and $[qx] = 0$ that $\dfrac{\partial x}{\partial p} = 0$, from (1) § 53;

hence $$[f(p, q), x] = \Sigma\left(\frac{\partial f}{\partial q}\frac{\partial x}{\partial p} - \frac{\partial f}{\partial p}\frac{\partial x}{\partial q}\right) = 0.$$

55. We now introduce the diagonal matrix

$$T \equiv \begin{pmatrix} T_1 & 0 & 0 & \dots \\ 0 & T_2 & 0 & \dots \\ 0 & 0 & T_3 & \dots \\ \dots & \dots & \dots & \end{pmatrix},$$

where $T_1, T_2, T_3, \dots$ are the Ritz terms of § 43.

Then if x is a quantum variable,

$$\begin{aligned}Tx - xT &= (\Sigma_k \{T(nk)\,x(km) - x(nk)\,T(km)\})\\ &= (T(nn)\,x(nm) - x(nm)\,T(mm)),\end{aligned}$$

since $T(nm) = 0$ if $n \neq m$.

[1] Q.M. I, equations (40).

$$\therefore \quad Tx - xT = (\{T_n - T_m\} x(nm))$$
$$= (\nu(nm) x(nm)),$$

since $T_n - T_m = \nu(nm)$, § 43, equation (3).

But $\dot{x} = (2\pi i \nu(nm) x(nm))$, § 52; hence

$$Tx - xT = \frac{1}{2\pi i}\dot{x} \quad \text{or} \quad xT - Tx = \frac{i}{2\pi}\dot{x}.$$

56. *The energy law and Bohr's frequency condition.*

From § 52 $$xH - Hx = \frac{ih}{2\pi}\dot{x}$$

and from § 55 $$xT - Tx = \frac{i}{2\pi}\dot{x},$$

therefore $$x(H - hT) - (H - hT)x = 0 \quad \text{.........(1).}$$

Writing H for x this becomes

$$H(H - hT) - (H - hT)H = 0$$

or $$HT - TH = 0 \quad \text{..................(2).}$$

But $xT - Tx = \frac{i}{2\pi}\dot{x}$, and writing H for x in this we have

$$HT - TH = \frac{i}{2\pi}\dot{H} \quad \text{..................(3).}$$

Hence from (2) and (3) $\dot{H} = 0$, and is constant. Therefore for non-degenerate systems, H is a diagonal matrix (§ 45 (iii)), whose terms may be written $H_1, H_2, H_3, \ldots$.

Again equation (1) gives

$$\Sigma_k x(nk)[H(km) - hT(km)] - \Sigma_k [H(nk) - hT(nk)] x(km) = 0$$

or $$x(nm)\{H_m - hT_m\} - \{H_n - hT_n\} x(nm) = 0.$$

$$\therefore \quad H_n - H_m = h(T_n - T_m) = h\nu(nm) \quad \text{......(4).}$$

This is Bohr's frequency condition that $h\nu$ is the energy change between two stationary states; it has been deduced by means of the new scheme of equations from the Ritz law of spectroscopy.

These equations were given by Born, Heisenberg, and Jordan[1], but the difficulties they met with in their efforts to preserve the Hamiltonian form of the equations of motion by a suitable form of matrix differentiation have been avoided by the use of Dirac's Poisson bracket theory, because in the classical mechanics the only differential coefficients essential to the theory can be put into Poisson bracket form.

Dirac's first great contribution to quantum mechanics was the discovery that the quantum conditions take a Poisson bracket form (§ 50), thus enabling the classical equations to be taken over into the quantum theory without a theory of formal matrix differentiation being necessary.

[1] Q.M. I and II.

CHAPTER X

THE HARMONIC OSCILLATOR

57. *The harmonic oscillator.*

The new theory of the harmonic oscillator was given by Heisenberg[1] in the paper in which he first outlined the new mechanics; he showed that the energy of a stationary state of the oscillator is $h\nu\,(n + \frac{1}{2})$, thus for the first time giving a dynamical theory which naturally led to the *half odd integer* quantum numbers which multiplet theory had so long asked for. Heisenberg's paper contained the ideas, but there was no attempt to give a rigorous theory of the oscillator; this was given later by Born and Jordan[2] as follows:

For the oscillator,

$$H = \frac{1}{2m_0} p^2 + \frac{1}{2} kq^2.$$

The Hamiltonian equations are $\dot{q} = \dfrac{p}{m_0}$, $\dot{p} = kq$.

$$\therefore\ \ddot{q} = -\frac{k}{\mu} q, \text{ and writing } \frac{k}{\mu} = (2\pi\nu_0)^2,$$

$$\ddot{q} + (2\pi\nu_0)^2\, q = 0 \quad \text{..................(1).}$$

Writing $q = (q\,(nm) \exp 2\pi i\nu\,(nm)\, t)$, we have from (1),

$$\{-\,4\pi^2\nu^2\,(nm) + 4\pi^2{\nu_0}^2\}\, q\,(nm) = 0$$

or

$$\{\nu^2\,(nm) - {\nu_0}^2\}\, q\,(nm) = 0 \quad \text{..............(2).}$$

The quantum condition

$$qp - pq = \frac{ih}{2\pi} 1$$

yields, (if we note that

$$p = m_0 \dot{q} \ \text{ or } \ p = (m_0 2\pi i\nu\,(nm)\, q\,(nm)),$$

[1] W. Heisenberg, Zs. f. Phys. **33**, p. 879, **1925**.
[2] Q.M. I.

and write down the diagonal term, the (nn) term of the matrix)

$$\Sigma_k \{q\,(nk)\, m_0\, 2\pi i\nu\,(kn)\, q\,(kn) - m_0\, 2\pi i \nu\,(nk)\, q\,(nk)\, q\,(kn)\} = \frac{ih}{2\pi}.$$

Since $\nu\,(nk) = -\,\nu\,(kn)$, and $q\,(kn) = q^*\,(nk)$ or

$$q\,(nk)\; q\,(kn) = |\, q\,(nk)\,|^2, \qquad (\S\ 43)$$

we have
$$-\,\Sigma_k 4 m_0 \pi i \nu\,(nk)\,|\, q\,(kn)\,|^2 = \frac{ih}{2\pi}.$$

$$\therefore\ \ \Sigma_k \nu\,(nk)\,|\, q\,(kn)\,|^2 = -\frac{h}{8\pi^2 m_0} \quad \text{.........(3).}$$

Also
$$H = \frac{1}{2} m_0 \dot{q}^2 + \frac{1}{2} m_0\,(2\pi\nu_0)^2 q^2$$

$$= (2\pi^2 m_0\, \Sigma_k \{-\,\nu\,(nk)\, q\,(nk)\, \nu\,(km)\, q\,(km) + \nu_0{}^2 q\,(nk)\, q\,(km)\}).$$

The energy $E\,(n)$ of the nth stationary state is $H\,(nn)$, for H is a diagonal matrix whose terms are the energies of the stationary states; therefore

$$E\,(n) = H\,(nn) = 2\pi^2 m_0 \Sigma_k \{\nu_0{}^2 + \nu^2\,(nk)\}\,|\, q\,(nk)\,|^2 \quad \text{...(4).}$$

Equations (2), (3), (4) contain the solution of the problem.

From (2), $q\,(nm) = 0$ except when $\nu\,(nm) = \pm\,\nu_0$, or when
$$E\,(n) - E\,(m) = \pm\, h\nu_0.$$

From (3) it follows that, corresponding to any n, an n' must exist for which $q\,(nn')$ is not zero, otherwise (3) would become $0 = -\dfrac{h}{8\pi^2 m_0}$; therefore, corresponding to n, an n' exists for which $E\,(n) - E\,(n')$ is either equal to $h\nu_0$ or to $-\,h\nu_0$.

Assuming there is no degeneracy (§ 45 (iii)), so that if $n \neq m$, then $E\,(n) \neq E\,(m)$, there are *at most* two values of m for which $E\,(n) - E\,(m) = \pm\, h\nu_0$.

Denoting these by n', n'', we have the possibilities

$$\left.\begin{aligned} E(n) - E(n') &= h\nu_0 \\ E(n) - E(n'') &= -h\nu_0 \end{aligned}\right\} \text{(A)},$$

or

$$\left.\begin{aligned} E(n) - E(n') &= -h\nu_0 \\ E(n) - E(n'') &= h\nu_0 \end{aligned}\right\} \text{(B)},$$

the former corresponding to

$$\left.\begin{aligned} \nu(nn') &= \nu_0 \\ \nu(nn'') &= -\nu_0 \end{aligned}\right\} \text{(A)},$$

and the latter to

$$\left.\begin{aligned} \nu(nn') &= -\nu_0 \\ \nu(nn'') &= \nu_0 \end{aligned}\right\} \text{(B)},$$

and in both cases $\nu(nn') = -\nu(nn'')$.

Hence (3) becomes, since $q(nk) = 0$ except when $k = n'$ or n'',

$$\nu(nn') \mid q(nn') \mid^2 + \nu(nn'') \mid q(nn'') \mid^2 = -\frac{h}{8\pi^2 m_0},$$

or
$$\nu(nn') \{\mid q(nn') \mid^2 - \mid q(nn'') \mid^2\} = -\frac{h}{8\pi^2 m_0} \quad \text{...(5)}.$$

Also (4) becomes

$$E(n) = 2\pi^2 m_0 [\{\nu_0^2 + \nu^2(nn')\} \mid q(nn') \mid^2 + \{\nu_0^2 + \nu^2(nn'')\} \mid q(nn'') \mid^2]$$

or
$$E(n) = 4\pi^2 m_0 \nu_0^2 \{\mid q(nn') \mid^2 + \mid q(nn'') \mid^2\} \quad \text{...(6)}.$$

We now consider the two cases given by equations (A) and (B).

It has been seen that for a given n there must be an n' and there may be an n''.

First suppose that n'' does not exist.

Then (5) and (6) become

$$\nu(nn') \mid q(nn') \mid^2 = -\frac{h}{8\pi^2 m_0} \quad \text{............(7)},$$

$$E(n) = 4\pi^2 m_0 \nu_0^2 \mid q(nn') \mid^2 \quad \text{............(8)},$$

(n'' being no longer a possible value of k in (3) and (4)).

For (7) to hold, $\nu(nn')$ must be negative; this rules out equations (A) and we can only have (B), where $\nu(nn') = -\nu_0$, so that $E(n') > E(n)$.

Equations (7) and (8) then become

$$|q(nn')|^2 = \frac{h}{8\pi^2 m_0 \nu_0}$$

and
$$E(n) = 4\pi^2 m_0 \nu_0^2 \left(\frac{h}{8\pi^2 m_0 \nu_0}\right),$$

or
$$E(n) = \frac{1}{2} h\nu_0.$$

Thus there is at most one n for which no n'' exists; let this be called n_0, so that

$$E(n_0) = \frac{1}{2} h\nu_0 \text{ and } |q(n_0 n_0')|^2 = \frac{h}{8\pi^2 m_0 \nu_0} \quad \ldots(9).$$

For all other n's there is both an n' and an n''.

Assuming for the moment that n_0 exists, we can find n_0' for which

$$\left.\begin{array}{l} E(n_0') - E(n_0) = h\nu_0, \\ \text{but there is no } n_0'' \text{ satisfying} \\ E(n_0'') - E(n_0) = -h\nu_0 \end{array}\right\} \quad \ldots\ldots\ldots\ldots(10).$$

Denote this n_0' by n_1. From this we can derive two new numbers n_1', n_1'', which both exist, satisfying

$$\left.\begin{array}{l} E(n_1') - E(n_1) = h\nu_0 \\ E(n_1'') - E(n_1) = -h\nu_0 \end{array}\right\}.$$

Comparing the second of these with the first of equations (10), namely,

$$E(n_0') - E(n_0) = h\nu_0 \text{ or } E(n_0) - E(n_1) = -h\nu_0$$

(since $n_1 = n_0'$), we see that n_1'' must be n_0.

Thus from n_1 we derive in this way n_1' and n_0.

If we denote n_1' by n_2, we can by a similar process from n_2 derive n_2' and n_1; then we denote n_2' by n_3 and proceed in the same way.

Thus we have the scheme:

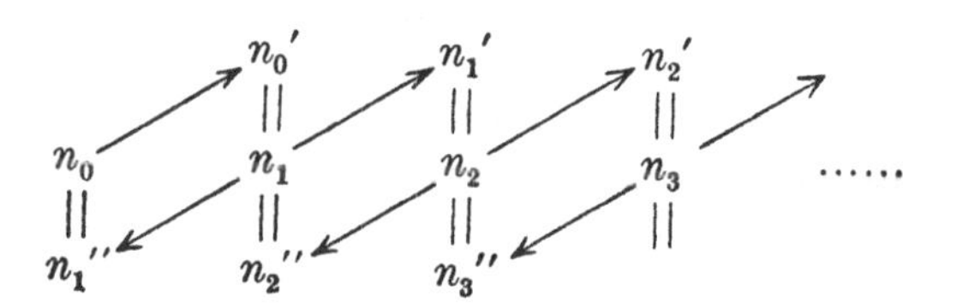

where $n'_{k-1} = n_k = n''_{k+1}$, except when $k = 0$, where only $n_0 = n_1''$; or what is the same, $n_k' = n_{k+1}$, $n_k'' = n_{k-1}$, except when $k = 0$, where only $n_0' = n_1$, there being no n_0''.

In this way is built up the sequence of numbers

$$n_0, n_1, n_2, \ldots,$$

bounded in the direction n_0, and such that $n_k' = n_{k+1}$, and $n_k'' = n_{k-1}$ $(k \neq 0)$.

Thus for $k \neq 0$, equations (5) and (6) become (since $\nu\,(nn')$ is now

$$\nu\,(n_k n_k') = \frac{E\,(n_k) - E\,(n_k')}{h} = -\,\nu_0),$$

$$|\,q\,(n_k n_{k+1})\,|^2 - |\,q\,(n_k n_{k-1})\,|^2 = \frac{h}{8\pi^2 m_0 \nu_0} \quad \ldots(11),$$

$$E\,(n_k) = 4\pi^2 m_0 \nu_0^2\,\{|\,q\,(n_k n_{k+1})\,|^2 + |\,q\,(n_k n_{k-1})\,|^2\} \quad (12).$$

Equations (11) are, writing $u_k \equiv |\,q\,(n_k n_{k+1})\,|^2$,

$$\left.\begin{aligned} u_k - u_{k-1} &= \frac{h}{8\pi^2 m_0 \nu_0} \\ u_{k-1} - u_{k-2} &= \frac{h}{8\pi^2 m_0 \nu_0} \\ &\cdots\cdots\cdots\cdots \\ &\cdots\cdots\cdots\cdots \\ u_1 - u_0 &= \frac{h}{8\pi^2 m_0 \nu_0} \end{aligned}\right\} \quad \ldots\ldots\ldots\ldots (13).$$

Also by equation (9),

$$u_0 = \frac{h}{8\pi^2 m_0 \nu_0}.$$

Adding, we have

$$u_k = \frac{h}{8\pi^2 m_0 \nu_0}(k+1)$$

or

$$| q(n_k n_{k+1}) |^2 = \frac{h(k+1)}{8\pi^2 m_0 \nu_0} \qquad \text{...........(14).}$$

Hence from (12) and (14) it follows that

$$E(n_k) = 4\pi^2 m_0 \nu_0^2 \frac{h}{8\pi^2 m_0 \nu_0}\{(k+1)+k\}$$

or

$$E(n_k) = h\nu_0\left(k+\frac{1}{2}\right) \qquad \text{..............(15).}$$

The whole of this depends upon the existence of this one n_0 for which there is an n_0' but no n_0''. If there is no such n_0 having this property, then we can start with any number N_0 and from it derive N_0' and N_0'', the former being denoted by N_1 and the latter by N_{-1}. We can then proceed as before and obtain a series of numbers ranging from $-\infty$ to ∞, as below:

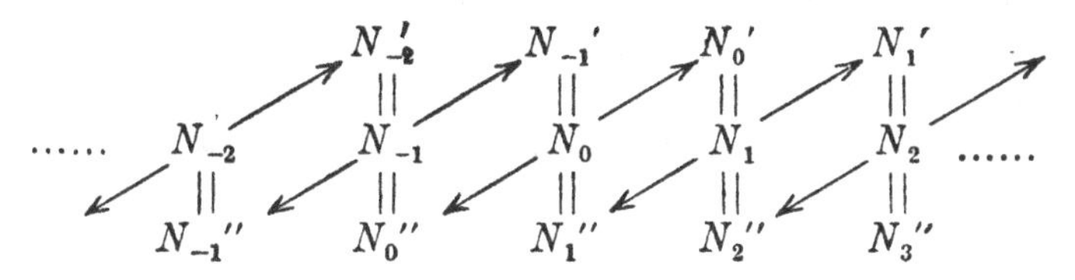

Thus equations (13) would hold for every k from $-\infty$ to ∞. But this is impossible since the magnitudes u_k form an arithmetical progression of *positive* terms, so that there must be a least one at which the series terminates in one direction; this corresponds to some finite k, which is contrary to the above result of assuming that there is no n_0. Hence an n_0 always exists.

Further, the sequence $n_0, n_1, n_2, \ldots$ contains every integer; for if an integer m were omitted, we could build up a sequence backwards and forwards from m in the above manner which must terminate on the left at some value m_0

for which $H\,(m_0 m_0) = \frac{1}{2} h\nu_0$. But $\frac{1}{2} h\nu_0 = H\,(n_0 n_0)$. Therefore m_0 must be n_0, so that m must be in the sequence $n_0, n_1, n_2, \ldots$.

Thus the solution of the problem is

$$q = \begin{pmatrix} 0, & q(01)\,e^{2\pi i\nu(01)t}, & 0, & 0 & \ldots \\ q(10)e^{2\pi i\nu(10)t}, & 0, & q(12)\,e^{2\pi i\nu(12)t}, & 0 & \ldots \\ 0, & q(21)\,e^{2\pi i\nu(21)t}, & 0, & q(23)\,e^{2\pi i\nu(23)t} & \ldots \\ \ldots & \ldots & \ldots & \ldots & \ldots \end{pmatrix},$$

where $\nu\,(n+1, n) = \nu_0$, $|\,q\,(n+1, n)\,|^2 = \dfrac{h}{8\pi^2 \mu\nu_0}\,(n+1)$,

$$E_n \equiv H\,(nn) = h\nu_0\left(n + \frac{1}{2}\right).$$

The value $H\,(00) \equiv \frac{1}{2} h\nu_0$, the 'nul-point energy' used by Planck and Nernst in the statistical problems of the quantum theory, is thus given naturally by the new theory. The *phase* in each component of the matrix is quite arbitrary, as $q\,(n+1, n) = |\,q\,(n+1, n)\,|\,e^{i\phi\,(n+1,\,n)}$, where $\phi\,(n+1, n)$ is an arbitrary phase constant, since only $|\,q\,(n+1, n)\,|$ is determined above, and not $q\,(n+1, n)$.

CHAPTER XI

THE CANONICAL TRANSFORMATION; PERTURBATION THEORY FOR NON-DEGENERATE AND DEGENERATE SYSTEMS

58. *The canonical transformation.*

$$P = SpS^{-1},\ Q = SqS^{-1}.$$

Consider a set of variables p, q satisfying the usual conditions

$$p_r p_s - p_s p_r = 0,\ q_r q_s - q_s q_r = 0,\ q_r p_s - p_s q_r = \frac{ih}{2\pi}\,\delta\,(rs)\,1,$$

where
$$\delta\,(rs) = 1,\ r = s,$$
$$= 0,\ r \neq s.$$

Transform them to P, Q by the transformation

$$P = SpS^{-1},\ Q = SqS^{-1},$$

where S is any matrix.

Then

$$\begin{aligned} Q_r P_s - P_s Q_r &= Sq_r S^{-1} Sp_s S^{-1} - Sp_s S^{-1} Sq_r S^{-1} \\ &= S\,(q_r p_s - p_s q_r)\,S^{-1},\ \text{since } S^{-1}S = 1 = SS^{-1} \\ &= S\,\frac{ih}{2\pi}\,\delta\,(rs)\,1S^{-1} \\ &= \frac{ih}{2\pi}\,\delta\,(rs)\,1, \end{aligned}$$

since S, S^{-1} commute with 1 and with any constant.

So
$$P_r P_s - P_s P_r = 0,$$
$$Q_r Q_s - Q_s Q_r = 0.$$

Thus the conditions of § 49 (iii) are satisfied and the transformation is canonical. These equations

$$P = SpS^{-1},\ Q = SqS^{-1}$$

are the analogue of a *contact transformation*[1] of the classical theory.

[1] Q.T.A. § 40.

It will next be shown that if $f(p, q)$ is a function of p, q developable in powers of p, q,

$$f(P, Q) = Sf(p, q)\, S^{-1},$$

the f being the same functional symbol on both sides.

It suffices to show that if this formula holds for two functions ϕ and ψ then it holds for $\phi + \psi$ and $\phi\psi$, as it is by the operations of addition and multiplication that f can be built up from p and q. The former is obvious, and the latter follows because if

$$\phi(P, Q) = S\phi(p, q)\, S^{-1}$$

and

$$\psi(P, Q) = S\psi(p, q)\, S^{-1},$$

then

$$\phi(P, Q)\,\psi(P, Q) = S\phi(p, q)\, S^{-1} S\psi(p, q)\, S^{-1}$$

$$= S\phi(p, q)\,\psi(p, q)\, S^{-1}.$$

Hence

$$f(P, Q) = Sf(p, q)\, S^{-1}.$$

In particular, if H is the Hamiltonian,

$$H(P, Q) = SH(p, q)\, S^{-1}.$$

59. *The condition* $S\tilde{S}^* = 1$.

This is the condition which S must satisfy in order to preserve the Hermite character of the matrices, which is essential if they are to represent real quantities (§ 44).

The symbol $\tilde{S}$ denotes the matrix whose rows are the columns and columns the rows of the matrix S; that is,

$$\tilde{S}(nm) = S(mn).$$

The asterisk denotes the conjugate complex quantity; S^* is the matrix whose components are the conjugates of those of S.

It may be noticed that if $a = bc$ then $\tilde{a} = \tilde{c}\tilde{b}$, for if $a = bc$, we have

$$a(nm) = \Sigma_k\, b(nk)\, c(km).$$

$$\therefore\ a(mn) = \Sigma_k\, c(kn)\, b(mk).$$

$$\therefore\ \tilde{a}(nm) = \Sigma_k\, \tilde{c}(nk)\, \tilde{b}(km).$$

$$\therefore\ \tilde{a} = \tilde{c}\tilde{b} \quad \text{..........................(1)}.$$

Now suppose the matrix p is of Hermite type and is transformed into the matrix P by the transformation $P = SpS^{-1}$; what is the condition that the new matrix P may be of Hermite type?

Since p is of Hermite type,

$$p(nm) = p^*(mn)$$

or

$$p = \tilde{p}^* \qquad \ldots\ldots\ldots\ldots\ldots\ldots\ldots(2).$$

By equation (1) it follows that since $P = SpS^{-1}$, then $\tilde{P} = \tilde{S}^{-1}\tilde{p}\tilde{S}$ and therefore

$$\tilde{P}^* = \tilde{S}^{-1*}\tilde{p}^*\tilde{S}^* = \tilde{S}^{-1*}p\tilde{S}^*, \text{ using (2)}.$$

If P is of Hermite type, $P = \tilde{P}^*$, so that we now have

$$P = \tilde{S}^{-1*}p\tilde{S}^*.$$

Comparing this with the original $P = SpS^{-1}$, we see that the Hermite character is preserved if

$$S^{-1} = \tilde{S}^* \qquad \ldots\ldots\ldots\ldots\ldots\ldots(3).$$

But $SS^{-1} = 1$, therefore

$$S\tilde{S}^* = 1 \qquad \ldots\ldots\ldots\ldots\ldots\ldots(4).$$

The condition (4) is equivalent to

$$\left.\begin{array}{l} \Sigma_k S(nk)\, S^*(mk) = 0 \\ \Sigma_k S(nk)\, S^*(nk) \;= 1 \end{array}\right\}$$

or the matrix S is 'orthogonal' (cf. § 68).

60. *Perturbation theory for one coordinate.*

The procedure is analogous to that of the classical theory[1]. The Hamiltonian $H(p, q)$ is given in the form

$$H_0(p, q) + \lambda H_1(p, q) + \lambda^2 H_2(p, q) + \ldots,$$

where λ is a constant indicating the scale of the perturbation. Suppose the original problem for which the Hamiltonian is $H_0(p, q)$ has been solved, so that p, q are known matrices and $H_0(p, q)$ is a diagonal matrix E_0.

We now seek for a transformation matrix S to new variables P, Q such that $P = SpS^{-1}$, $Q = SqS^{-1}$ and

[1] Q.T.A. chap. XXI.

therefore $H(P, Q) = SH(p, q) S^{-1}$ (§ 58), with the condition $S\tilde{S}^* = 1$ to preserve the Hermite type of matrix (§ 59).

If the new $H(P, Q)$ is a diagonal matrix E, so that

$$SH(pq) S^{-1} = E \qquad \text{..............(1)},$$

the problem is solved, and the components of E are the energy levels of the perturbed system.

Equation (1) is the analogue of the Hamilton-Jacobi equation of the classical theory and the matrix S that of the S-function of that theory[1].

First approximation.

We have $H = H_0 + \lambda H_1 + \ldots$ and we assume that

$$\left. \begin{aligned} E &= E_0 + \lambda E_1 + \ldots \\ S &= S_0 + \lambda S_1 + \ldots \end{aligned} \right\} \qquad \text{...............(2)},$$

where the E's and H_0 are diagonal matrices.

The condition (1) viz. $SHS^{-1} = E$ becomes, on multiplying *behind* by S on each side, $SH = ES$.

Substituting in this from (2), we have

$$(S_0 + \lambda S_1 \ldots)(H_0 + \lambda H_1 \ldots) = (E_0 + \lambda E_1 \ldots)(S_0 + \lambda S_1 \ldots).$$

Equating coefficients of powers of λ, we have to a first approximation

$$\left. \begin{aligned} S_0 H_0 &= E_0 S_0 \\ S_1 H_0 + S_0 H_1 &= E_0 S_1 + E_1 S_0 \end{aligned} \right\} .$$

Since $\qquad H_0 = E_0,$

$$\therefore \left. \begin{aligned} S_0 H_0 - H_0 S_0 &= 0 \\ S_1 H_0 - H_0 S_1 &= E_1 S_0 - S_0 H_1 \end{aligned} \right\} \qquad \text{...........(3)}.$$

Two cases arise, according as to whether the original system is degenerate or not.

61. *Undisturbed system non-degenerate.*

This is a system where no frequencies $\nu_0(nm)$ are zero when $n \neq m$ (cf. § 45 (iii)); that is, no two energy levels

[1] Q.T.A. chap. VI, §§ 40–42.

$H_0(nn)$, $H_0(mm)$ are the same; all the terms of the diagonal matrix H_0 are different.

The first equation of (3) § 60 is

$$\Sigma_k \{S_0(nk) H_0(km) - H_0(nk) S_0(km)\} = 0,$$

or

$$S_0(nm)[H_0(mm) - H_0(nn)] = 0,$$

since

$$H_0(nm) = 0, n \neq m.$$

$$\therefore \quad S_0(nm) h\nu_0(mn) = 0 \quad \ldots\ldots\ldots\ldots\ldots(1).$$

Since the original system is non-degenerate, $\nu_0(mn) \neq 0$ if $m \neq n$, and therefore $S_0(nm) = 0$ if $m \neq n$, but $S_0(nn)$ is not determinate from (1).

Thus S_0 is a diagonal matrix.

But since

$$S\tilde{S}^* = 1,$$

$$(S_0 + \lambda S_1)(\tilde{S}_0^* + \lambda \tilde{S}_1^*) = 1.$$

$$\therefore \quad S_0\tilde{S}_0^* = 1, \text{ and } S_1\tilde{S}_0^* + S_0\tilde{S}_1^* = 0.$$

From the former, since S_0 is a diagonal matrix

$$S_0(nn) S_0^*(nn) = 1.$$

$$\therefore \quad |S_0(nn)|^2 = 1 \quad \ldots\ldots\ldots\ldots\ldots(2).$$

Thus the 'phase' of $S_0(nn)$ is indeterminate and if we write

$$S_0(nn) = |S_0(nn)| e^{i\phi(nn)}$$
$$= e^{i\phi(nn)} \text{ from (2),}$$

we can absorb $e^{i\phi(nn)}$ into the usual unwritten time factor, and write $S_0(nn) = 1$. So we have effectively $S_0 = 1$.

The second equation of (3) § 60 is now

$$S_1H_0 - H_0S_1 = E_1 - H_1.$$

$$\therefore \quad \Sigma_k \{S_1(nk) H_0(km) - H_0(nk) S_1(km)\}$$
$$= E_1(nm) - H_1(nm),$$

or $\quad S_1(nm)[H_0(mm) - H_0(nn)] = E_1(nm) - H_1(nm),$

or

$$-S_1(nm) h\nu_0(nm) = E_1(nm) - H_1(nm).$$

If $n \neq m$, then

$$S_1(nm) h\nu_0(nm) = H_1(nm),$$

since E_1 is a diagonal matrix; and if $n = m$, then since $\nu_0(nn) = 0$,

$$E_1(nn) = H_1(nn).$$

Thus H_1 is not a diagonal matrix, but its diagonal terms are those of E_1. If we use the symbol $\overline{H}_1$ to denote the 'time mean' of the matrix H_1 over the undisturbed path, i.e. the matrix formed by writing all the terms of H_1 zero except the constant diagonal terms, then $E_1 = \overline{H}_1$ and $E = H_0 + \lambda \overline{H}_1$ give the new energy levels. This is exactly the analogue of the classical theory where the addition to the energy is the time mean of the Hamiltonian perturbation term taken over the undisturbed path[1].

62. *Undisturbed system degenerate.*

Suppose, for instance, that

$$H_0(n_1 n_1) = H_0(n_2 n_2) = H_0(n_3 n_3),$$

where $n_1 \neq n_2 \neq n_3$, so that three energy levels of the original system fall together. Then $\nu_0(nm) = 0$, not only if $m = n$, but also if m, n are any pair of n_1, n_2, n_3.

Then from equation (1) § 61, $S_0(nm)\,\nu_0(nm) = 0$, so that as before $S_0(nn)$ is not zero; but as not before, $S_0(nm)$ is not zero if n, m are any pair of n_1, n_2, n_3. In other cases $S_0(nm) = 0$. Thus S_0 is a matrix made up of constant terms but is not a diagonal matrix; the pattern of its matrix is given on the next page.

Also $S_0 \tilde{S}_0^* = 1$ shows that $S_0(nn) = 1$, except for $n = n_1$, n_2 or n_3, when its value is not unity. The crosses in the figure denote terms not equal to unity; the empty squares are occupied by zeros.

From the second equation of (3) § 60 we have

$$S_1 H_0 - H_0 S_1 = E_1 S_0 - S_0 H_1.$$

$$\therefore \quad -S_1(nm)\,\nu_0(nm) = E_1(nn)\,S_0(nm) - \Sigma_k S_0(nk)\,H_1(km) \quad \text{......(1)}.$$

[1] Q.T.A. chap. xx, §§ 133, 139.

If neither of n, m are members of the set n_1, n_2, n_3,

$$-S_1(nm)\,\nu_0(nm) = -S_0(nn)\,H_1(nm), \text{ since } S_0(nm) = 0$$

then. $\therefore S_1(nm)\,\nu_0(nm) = H_1(nm)$.

Thus $H_1(nm)$ has a value, not zero.

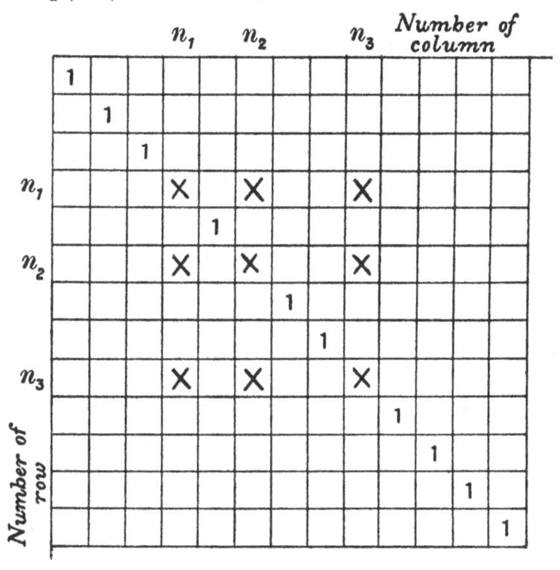

Also if n is not one of n_1, n_2, n_3, we have, putting $n = m$ in (1) and using $S_0(nn) = 1$,

$$0 = E_1(nn) - H_1(nn)$$

or

$$E_1(nn) = H_1(nn).$$

Thus the diagonal terms of H_1 still give those of E_1, except in the case where $n = n_1$ or n_2 or n_3.

Thus E_1 is completely determined, except for three of its terms. To find these, put $n = n_1$ in (1). Then

$$\begin{aligned} -S_1(n_1m)\,\nu_0(n_1m) &= E_1(n_1n_1)\,S_0(n_1m) \\ &\quad - [S_0(n_1n_1)\,H_1(n_1m) + S_0(n_1n_2)\,H_1(n_2m) \\ &\quad + S_0(n_1n_3)\,H_1(n_3m)] = 0, \end{aligned}$$

since $S_0(n_1 k) = 0$ except for $k = n_1$, n_2 or n_3 (cf. the matrix pattern given above).

Now put m successively n_1, n_2, n_3 and note that $\nu_0(n_2 n_3)$, etc. are zero.

We then have

$$0 = S_0(n_1 n_1)[-E_1(n_1 n_1) + H_1(n_1 n_1)] + S_0(n_1 n_2) H_1(n_2 n_1) + S_0(n_1 n_3) H_1(n_3 n_1),$$

$$0 = S_0(n_1 n_1) H_1(n_1 n_2) + S_0(n_1 n_2)[-E_1(n_1 n_1) + H_1(n_2 n_2)] + S_0(n_1 n_3) H_1(n_3 n_2),$$

$$0 = S_0(n_1 n_1) H_1(n_1 n_3) + S_0(n_1 n_2) H_1(n_2 n_3) + S_0(n_1 n_3)[-E_1(n_1 n_1) + H_1(n_3 n_3)].$$

Eliminating $S_0(n_1 n_1)$, $S_0(n_1 n_2)$ and $S_0(n_1 n_3)$, and writing $E_1 = E_1(n_1 n_1)$ we have

$$0 = \begin{vmatrix} H_1(n_1 n_1) - E_1, & H_1(n_2 n_1), & H_1(n_3 n_1) \\ H_1(n_1 n_2), & H_1(n_2 n_2) - E_1, & H_1(n_3 n_2) \\ H_1(n_1 n_3), & H_1(n_2 n_3), & H_1(n_3 n_3) - E_1 \end{vmatrix} \quad \ldots(2).$$

The same equation would have been found for $E_1(n_2 n_2)$ and $E_1(n_3 n_3)$ had we put $n = n_2$ or n_3 in (1) and proceeded as above, so that the three missing terms of E_1 are given as the roots of the cubic (2). Thus the energy levels of the perturbed system are found, the three coincident levels of H_0 splitting up into three new and different levels.

It may be observed that the matrix H_1 has constant terms not only in its diagonal but at the crossing points of the rows n_1, n_2, n_3 and the columns n_1, n_2, n_3. Thus its 'time mean' $\overline{H}_1$ would not be a diagonal matrix but would be of the pattern of S_0 (p. 91), so that it fails to yield all the energy levels as it was able to in the non-degenerate case.

But (2) shows that the levels are given by

$$S_0 \overline{H}_1 S_0^{-1} = E_1,$$

for this equation is

$$S_0 \overline{H}_1 = E_1 S_0.$$

Since $\overline{H}_1$ has only the terms $H_1(nn)$, $H_1(n_r n_s)$ where n_r, n_s are a pair of n_1, n_2, n_3, this equation yields

$$\Sigma S_0(n_r n_s) H_1(n_s n_s) = E_1(n_r n_r) S_0(n_r n_s) \quad \text{......(3)}$$

and the elimination of the S_0's from the three equations of the type (3) gives the same cubic for the E_1's as before, namely equation (2).

63. *Summary of the procedure.*

We take the time mean of the disturbing function λH_1 for the undisturbed path. If the original system is not degenerate, this time mean $\overline{\lambda H_1}$ is a diagonal matrix and the terms of $H_0 + \overline{\lambda H_1}$ are the energy levels sought for. If the original system is degenerate, then $\overline{\lambda H_1}$ is a matrix of constant terms but not a diagonal matrix. If a transformation $S\overline{\lambda H_1}S^{-1}$ is carried out so that this becomes a diagonal matrix λE_1, then the terms of $E_0 + \lambda E_1$ are the energy levels.

The procedure is exactly analogous to that of the classical theory as applied in the older quantum theory[1].

64. *Perturbation theory to a second approximation for a non-degenerate system.*

Using the first approximation results of §§ 60, 61 we have

$$SHS^{-1} = E.$$

$$\therefore \; SH = ES \quad \text{....................(1).}$$

Further, $S = 1 + \lambda S_1 + \lambda^2 S_2 + \ldots$, since $S_0 = 1$ for a non-degenerate system. Writing

$$H = H_0 + \lambda H_1 + \lambda^2 H_2 + \ldots,$$

$$E = E_0 + \lambda E_1 + \lambda^2 E_2 + \ldots,$$

and substituting in (1) we have

$$(1 + \lambda S_1 + \lambda^2 S_2 + \ldots)(H_0 + \lambda H_1 + \lambda^2 H_2 + \ldots)$$
$$= (E_0 + \lambda E_1 + \lambda^2 E_2 + \ldots)(1 + \lambda S_1 + \lambda^2 S_2 + \ldots).$$

[1] Q.T.A. §§ 139, 143.

Equating to zero the coefficients of the powers of λ, we obtain

$$\left.\begin{aligned} H_0 &= E_0 \\ S_1 H_0 + H_1 &= E_1 \\ S_2 H_0 + S_1 H_1 + H_2 &= E_0 S_2 + E_1 S_1 + E_2 \end{aligned}\right\}.$$

For the first order terms, we have from § 61

$$E_1(nn) = H_1(nn)$$

and $$S_1(nm)\, h\nu_0(nm) = H_1(nm),\ (n \neq m) \quad \ldots\ldots(2).$$

The second order terms yield

$$S_2 H_0 - H_0 S_2 + S_1 H_1 + H_2 - E_1 S_1 = E_2.$$

$$\therefore\quad - S_2(nm)\, h\nu_0(nm) + \Sigma_k S_1(nk)\, H_1(km) + H_2(nm) - E_1(nn)\, S_1(nm) = E_2(nm).$$

Writing $m = n$, we have

$$\Sigma_k S_1(nk)\, H_1(kn) + H_2(nn) - E_1(nn)\, S_1(nn) = E_2(nn) \ldots\ldots(3).$$

The condition $S\tilde{S}^* = 1$ yielded the first result $S_0 = 1$ and also the equation

$$S_1 \tilde{S}_0^* + S_0 \tilde{S}_1^* = 0 \qquad (\S\ 61),$$

or $$S_1 + \tilde{S}_1^* = 0$$

or $$S_1(nn) + S_1^*(nn) = 0.$$

Writing $S_1(nn) = Ce^{i\phi}$, where C is real and ϕ is the phase, we have

$$Ce^{i\phi} + Ce^{-i\phi} = 0$$

or $$2C \cos\phi = 0.$$

The phase is again arbitrary and $C = 0$. Therefore

$$S_1(nn) = 0 \quad \ldots\ldots\ldots\ldots\ldots\ldots\ldots(4).$$

Hence equation (3) becomes, using (4) and (2),

$$E_2(nn) = H_2(nn) + \Sigma_k{}' \frac{H_1(nk)\, H_1(kn)}{h\nu_0(nk)} \quad \ldots\ldots(5),$$

where the dash denotes that the term $k = n$ is omitted in the summation.

Thus the energy is found to the second order of approximation.

65. Heisenberg[1] has expressed the view that though we are far from a rigorous theory of convergence of quantum theory magnitudes, there is this difference between classical and quantum theory that, for example in the last equation (5) of § 64, the $\nu_0\ (nk)$ of the denominator cannot be made $<$ the ϵ of analysis; there is a *least* $\nu_0\ (nk)$ due to the n series and the k series being bounded in one direction and heaping up to a limit as in the hydrogen spectrum, and not going to infinity both ways at equidistant intervals, as in the classical theory. Thus the convergence questions of the classical perturbation theory which give rise to the notorious difficulties of the problem of three bodies probably do not arise in the corresponding quantum theory problem.

[1] W. Heisenberg, Math. Annalen, **95**, p. 683, **1926**.

CHAPTER XII

THE ANHARMONIC OSCILLATOR; INTENSITIES OF SPECTRAL LINES; MATRICES AND QUADRATIC FORMS

66. *The anharmonic oscillator.*

This is worked out to illustrate the theory of perturbations of §§ 61–64; the importance of the result in connection with the theory of band spectra[1] is well known.

The Hamiltonian for the anharmonic oscillator is

$$H = \underbrace{\frac{1}{2m_0}p^2 + \frac{1}{2}m_0(2\pi\nu_0)^2 q^2}_{H_0(p,q)} + \underbrace{\lambda q^3}_{\lambda H_1(p,q)} + \ldots, \text{ where } \lambda \text{ is small.}$$

$H_0(p, q)$ is the Hamiltonian for the harmonic oscillator, which is perturbed by a term λq^3 in the Hamiltonian.

The solution for the original problem (the harmonic oscillator) has been found (§ 57) and is $q = (q(nm))$, where $q(nm) = 0$ except when

$$m = n \pm 1, \text{ and } |q(n, n-1)|^2 = \frac{nh}{8\pi^2 m_0 \nu_0} = Cn,$$

suppose, where $C = \dfrac{h}{8\pi^2 m_0 \nu_0}$.

Also H_0 is a diagonal matrix and

$$H_0(nn) = h\nu_0\left(n + \frac{1}{2}\right);$$

further,

$$\nu(n, n-1) = \nu_0 \text{ and } \nu(n, n+1) = -\nu_0.$$

Again

$$\nu(n, k) = \{H_0(nn) - H_0(kk)\}/h$$
$$= \left\{h\nu_0\left(n + \frac{1}{2}\right) - h\nu_0\left(k + \frac{1}{2}\right)\right\}\Big/h.$$

$$\therefore\ \nu(n, k) = (n - k)\nu_0.$$

(Statements from "The solution for the original problem" to here are bracketed together as (1).)

[1] Q.T.A. chap. XVII.

To calculate the perturbation, we have $H_1 = q^3$.

$$\therefore \quad H_1 (nm) = \Sigma_k \Sigma_l q (nk) q (kl) q (lm).$$

Since $q (ns) = 0$ except when s is either $n - 1$ or $n + 1$, the only possible values of k, l, m, for a given n are given by the scheme:

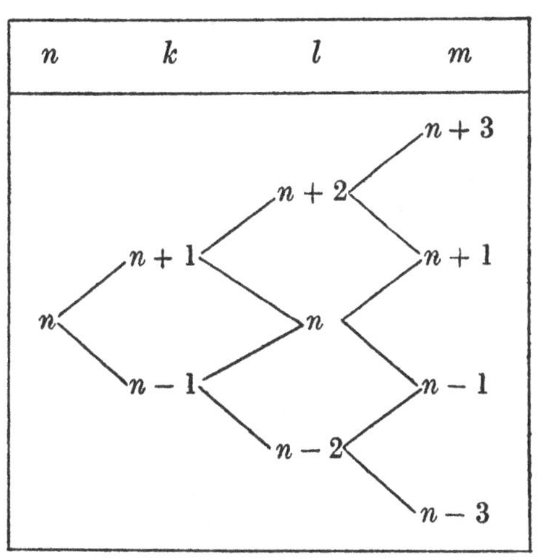

Thus $H_1 (nm) = 0$ except when $m = n - 3, n - 1, n + 1$ or $n + 3$.

The non-zero values of $H_1 (nm)$ are:

(i) $H_1 (n, n - 3)$.

The scheme shows the only possibilities to be

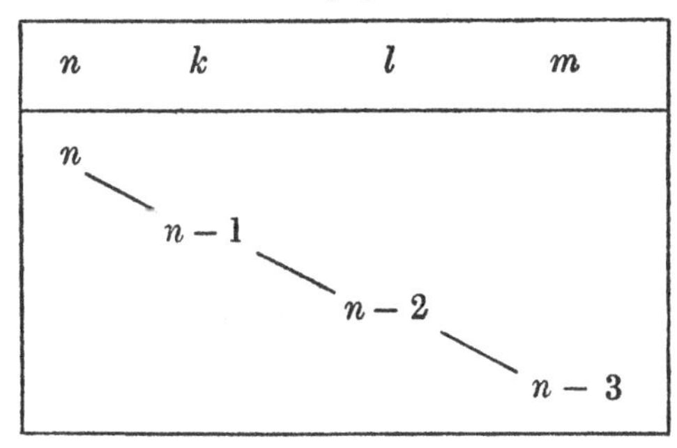

$$\therefore \quad H_1(n, n-3) = \Sigma_k \Sigma_l q(nk)\, q(kl)\, q(lm)$$
$$= q(n, n-1)\, q(n-1, n-2)\, q(n-2, n-3).$$

In the general result for $E_2(nn)$, which it is our aim to find, we require $H_1(nk)\, H_1(kn)$ (§ 64, equation (5)), so that we calculate $H_1(n, n-3)\, H_1(n-3, n)$, which is equal to

$$q(n, n-1)\, q(n-1, n-2)\, q(n-2, n-3)$$
$$\times\, q(n-3, n-2)\, q(n-2, n-1)\, q(n-1, n),$$

which

$$= |q(n, n-1)|^2\, |q(n-1, n-2)|^2\, |q(n-2, n-3)|^2,$$

since the matrices are of the Hermite type,

$$= C^3 n\, (n-1)\, (n-2),$$

from the data of (1).

(ii) $H_1(n, n+3)$ in like manner yields

$$H_1(n, n+3)\, H_1(n+3, n)$$
$$= |q(n, n+1)|^2\, |q(n+1, n+2)|^2\, |q(n+2, n+3)|^2$$
$$= C^3(n+1)\,(n+2)\,(n+3).$$

(iii) $H_1(n, n-1)$.

The scheme shows the possibilities:

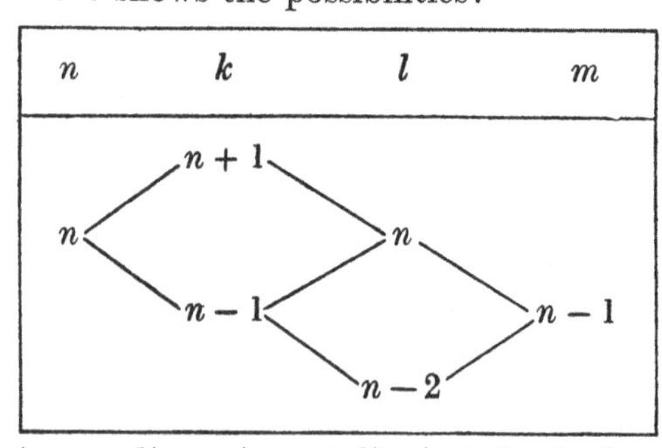

$$\therefore \quad H_1(n, n-1) = q(n, n+1)\, q(n+1, n)\, q(n, n-1)$$
$$+\, q(n, n-1)\, \{q(n-1, n)\, q(n, n-1)$$
$$+\, q(n-1, n-2)\, q(n-2, n-1)\}$$
$$= q(n, n-1)\, \{|q(n+1, n)|^2 + |q(n, n-1)|^2$$
$$+\, |q(n-1, n-2)|^2\}$$
$$= q(n, n-1)\, C\,(n+1+n+n-1)$$
$$= 3Cn\, q(n, n-1).$$

$$\therefore \quad H_1(n, n-1) H_1(n-1, n) = 9C^2 n^2 |q(n, n-1)|^2$$
$$= 9C^3 n^3.$$

(iv) $H_1(n, n+1)$ in like manner yields

$$H_1(n, n+1) H_1(n+1, n) = 9C^3(n+1)^3.$$

Hence substituting in

$$E_2(nn) = H_2(nn) + \Sigma_k{}' \frac{H_1(nk) H_1(kn)}{h\nu_0(nk)} \quad (\S\ 64),$$

and since H_2 is absent from H, to this order, we have

$$E_2(nn) = \frac{C^3 n(n-1)(n-2)}{h\nu(n, n-3)} + \frac{C^3(n+1)(n+2)(n+3)}{h\nu(n, n+3)}$$
$$+ \frac{9C^3 n^3}{h\nu(n, n-1)} + \frac{9C^3(n+1)^3}{h\nu(n, n+1)},$$

and since $\nu(n, k) = (n-k)\nu_0$, from the data (1),

$$E_2(nn) = \frac{C^3}{h\nu_0}\left[\frac{n(n-1)(n-2)}{3} - \frac{(n+1)(n+2)(n+3)}{3} + 9n^3 - 9(n+1)^3\right]$$

$$= \frac{C^3}{h\nu_0}\{-3n^2 - 3n - 2 - 9(3n^2 + 3n + 1)\}$$

$$= -\frac{C^3}{h\nu_0}(30n^2 + 30n + 11).$$

Also $E_1(nn) = H_1(nn) = 0$, since $H_1(nm) = 0$ except when $m = n-3$, $n-1$, $n+1$ or $n+3$; and

$$E_0(nn) = H_0(nn) = h\nu_0\left(n + \frac{1}{2}\right).$$

Hence the energy of the nth state of the anharmonic oscillator

$$= E_0(nn) + \lambda E_1(nn) + \lambda^2 E_2(nn) + \ldots$$

$$= h\nu_0\left(n + \frac{1}{2}\right) - \lambda^2\left(\frac{C^3}{h\nu_0}\right)(30n^2 + 30n + 11)$$

$$= h\nu_0\left(n + \frac{1}{2}\right) - \frac{15\lambda^2 h^2}{32\nu_0(2\pi^2 m_0 \nu_0)^3}\left(n^2 + n + \frac{11}{30}\right),$$

inserting the value of C from (1) (cf. Q.M. I, p. 883).

The older quantum theory gave the result[1]

$$h\nu_0 n - \frac{15\lambda^2 h^2}{32\nu_0 (2\pi^2 m_0 \nu_0)^3} \cdot n^2.$$

67. *Intensities and frequencies of spectral lines.*

On the classical theory an oscillator whose electric moment is the real part of $\mathbf{C}e^{2\pi i\nu t}$ radiates energy per second of amount

$$\frac{(2\pi\nu)^4}{3c^3}(\mathbf{CC}^*), \text{ (where } \mathbf{C}^* \text{ is the vector conjugate to } \mathbf{C}),$$

or

$$\frac{(2\pi\nu)^4}{3c^3}|\mathbf{C}|^2.$$

On the new quantum theory the electric moment is represented by a matrix whose components are

$$\mathbf{A}(nm)\, e^{2\pi i\nu(nm)t},$$

each component representing a transition. If $a(nm)$ is the Einstein probability coefficient for the transition $n \to m$, i.e. the number of transitions per second, the energy emission per second $= a(nm)\, h\nu(nm)$, since $h\nu(nm)$ is the emission in a transition.

$$\therefore \quad a(nm)\, h\nu(nm) = \frac{\{2\pi\nu(nm)\}^4}{3c^3}|\mathbf{A}(nm)|^2 \quad \text{...(1).}$$

The Einstein coefficient is the measure of the intensity of the spectral line whose frequency is $\nu(nm)$ and is found by formula (1) directly from the amplitude $\mathbf{A}(nm)$ given by the matrix representing the electric moment.

68. *Matrices and quadratic forms.*

Consider a quadratic form $A(xx) \equiv \Sigma_{nm}\, a(nm)\, x_n x_m$ and the matrix $a \equiv (a(nm))$ associated with it.

Make a linear transformation of the variables x to new variables y, where $x_n = \Sigma_l\, S(ln)\, y_l$, and let $S \equiv (S(nm))$ be the matrix associated with this. Let the transformation be also 'orthogonal,' i.e. such that the quadratic 'unit'

[1] Q.T.A. p. 216.

form $U(xx) \equiv \Sigma_n x_n^2$, with its 'unit' matrix 1, transforms into $U(yy) \equiv \Sigma_n y_n^2$.

Let $A(xx)$ become $B(yy) \equiv \Sigma_{nm} b(nm) y_n y_m$, with its matrix $b \equiv (b(nm))$.

Carry out the substitution of the y's for the x's in $A(xx)$. Then

$$\begin{aligned} B(yy) &= \Sigma_{nm} a(nm) \Sigma_l S(ln) y_l \Sigma_k S(km) y_k \\ &= \Sigma_{nm} \Sigma_l \Sigma_k S(ln) a(nm) S(km) y_l y_k. \end{aligned}$$

Changing l into n, n to k, m to l, and k to m, this becomes

$$B(yy) = \Sigma_{kl} \Sigma_n \Sigma_m S(nk) a(kl) S(ml) y_n y_m.$$

Hence
$$\begin{aligned} b(nm) &= \Sigma_{kl} S(nk) a(kl) S(ml) \\ &= \Sigma_{kl} S(nk) a(kl) \tilde{S}(lm), \end{aligned}$$
or
$$b = Sa\tilde{S}.$$

The quadratic unit form $U(xx)$ with its unit matrix transforms into $U(yy)$ with its unit matrix; therefore

$$1 = S1\tilde{S} \quad \text{or} \quad \tilde{S} = S^{-1}.$$

$$\therefore \quad b = SaS^{-1}.$$

The new quadratic form is $\Sigma_{nm} b(nm) y_n y_m$ and if it is a sum of squares, b is a diagonal matrix. Thus the problem of transforming a matrix a by the use of a transformation matrix S to a diagonal matrix b is the problem of reducing a quadratic form to a sum of squares.

This problem arose in § 60 where SHS^{-1} had to be transformed to a diagonal matrix E.

For *finite* quadratic forms[1], it is well known that the quadratic form $\Sigma a(nm) x_n x_m$ transforms to a sum of squares $\Sigma b(nn) y_n^2$ if the $b(nn)$'s are the roots of the equation in b:

$$\begin{vmatrix} a(11) - b, & a(12), & \dots & a(1s) \\ a(21), & a(22) - b, & \dots & a(2s) \\ \dots & \dots & & \dots \\ a(s1), & a(s2), & \dots & a(ss) - b \end{vmatrix} = 0.$$

[1] Courant and Hilbert, chap. I, § 3.

The terms of the diagonal matrix b are found without of necessity finding S. If the matrix a were H, these would be the energies of the stationary states; they are the 'eigenwerte' of the quadratic form $H(xx)$ corresponding to the matrix H.

But, as we know, the matrices (and the associated quadratic forms) for quantum variables are not finite (§ 47); the corresponding theory has only been so far developed for a limited class of forms by Hilbert and Hellinger[1]. Though the forms here used are more general, yet the properties of this limited class of forms throw light on this problem. These forms have the property that $\Sigma_{nm} H(nm)\, x_n x_m$ does not transform into $\Sigma_n E(nn)\, y_n^2$, where there is a discrete set of n's, but into

$$\Sigma_n E(nn)\, y_n^2 + \int E(\phi)\,[y(\phi)]^2\, d\phi,$$

the first set of terms corresponding to discrete n's and the latter integral to a continuous range of ϕ. These correspond physically to a line spectrum and a continuous spectrum[2].

This outlook will not be developed further here, as the Schrödinger theory (chapter XVIII), with its more highly developed analysis, leads to these results.

69. *Matrix notation in general.*

When several quantum numbers define a state, the letter n can be used to represent the state $(n_1, n_2, \ldots n_s)$, and the term (nm) of a matrix corresponds to a transition from the state $(n_1 \ldots n_s)$ to the state $(m_1 \ldots m_s)$.

[1] E. Hellinger, Crelle, **136**, p. 1, **1910**.
[2] Q.M. II, pp. 590–5.

CHAPTER XIII

ANGULAR MOMENTUM RELATIONS; SELECTION AND POLARISATION RULES

70. *Angular momentum relations for one electron* (*Dirac*[1]). Let x, y, z, p_x, p_y, p_z, be the Cartesian coordinates and momenta of an electron. Then the components of angular momentum are m_x, m_y, m_z, where $m_x = yp_z - zp_y$, etc.

The following are a series of angular momentum properties required in the theory of the general Zeeman effect.

(i) $[m_x x] = 0$, $[m_x y] = z$, $[m_x z] = -y$.

The general quantum conditions of § 51 here run:

$$[yz] = 0, \text{ etc., } [p_y p_z] = 0, \text{ etc., } [yp_z] = 0, \text{ etc., } [xp_x] = 1, \text{ etc.} \quad \ldots\ldots(1).$$

Hence

$$\begin{aligned} [m_x y] &= [yp_z - zp_y, y] \\ &= [yp_z, y] - [zp_y, y] \\ &= [yy]\, p_z + y\, [p_z y] - [zy]\, p_y - z\, [p_y y] \\ &= z\, [yp_y] = z, \text{ using (1).} \end{aligned}$$

$$\therefore \quad [m_x y] = z.$$

So

$$[m_x z] = -y,$$
$$[m_x x] = 0.$$

These formulae hold after cyclic change of x, y, z.

(Note that the + sign occurs on the right-hand side if the cyclic order $xyzx$ is present in the formula and the − sign if the reverse order $xzyx$ is present.)

(ii) $[m_x p_x] = 0$, $[m_x p_y] = p_z$, $[m_x p_z] = -p_y$.

These are proved in the same way as (i).

[1] P. A. M. Dirac, Proc. Roy. Soc. A. **111**, p. 281, March **1926**.

(iii) $[m_y m_z] = m_x$, $[m_z m_x] = m_y$, $[m_x m_y] = m_z$.

For

$$\begin{aligned}[m_y m_z] &= [zp_x - xp_z, m_z] \\ &= [zm_z]\, p_x + z\,[p_x m_z] - [xm_z]\, p_z - x\,[p_z m_z] \\ &= -zp_y + yp_z, \text{ using (i) and (ii)} \\ &= m_x.\end{aligned}$$

$$\therefore\ [m_y m_z] = m_x,$$

and so on, for cyclic change of x, y, z.

71. *Angular momentum relations for several electrons (Dirac*[1]*; Heisenberg and Jordan*[2]*).*

For all the electrons of a system, the components of angular momentum are M_x, M_y, M_z, where $M_x = \Sigma m_x$, etc.

(i) Any function of the coordinates and momenta of one electron (r) commutes with any function of those of another (s) on account of the general relations

$$[p_r p_s] = 0, \quad [q_r q_s] = 0, \quad [p_r q_s] = 0$$

and the theorem of § 54.

(ii) For an electron (r),

$$\begin{aligned}[M_x y_r] &= [\Sigma m_x, y_r] \\ &= [m_{1x} + m_{2x} + m_{3x} + \ldots + m_{rx} + \ldots, y_r] \\ &= [m_{rx}, y_r], \text{ using (i) § 71} \\ &= z_r, \text{ using (i) § 70,}\end{aligned}$$

or dropping the suffix r

$$[M_x y] = z, \text{ for each electron.}$$

So $$[M_x z] = -y$$

and $$[M_x x] = 0, \text{ for each electron.}$$

(iii) In like manner

$$[M_x p_x] = 0, \quad [M_x p_y] = p_z, \quad [M_x p_z] = -p_y$$

for each electron of the system.

[1] P. A. M. Dirac, Proc. Roy. Soc. A. **111**, p. 281, **1926**.
[2] Q.M. II.

(iv) $[M_y M_z] = M_x$, etc.

For $$[M_y M_z] = [\Sigma_k m_{ky}, \Sigma_l m_{lz}]$$
$$= \Sigma_k \Sigma_l [m_{ky} m_{lz}].$$

But $$m_{ky} m_{lz} = 0,\ k \neq l \text{ by (i) § 71.}$$

$$\therefore\ [M_y M_z] = \Sigma_k [m_{ky} m_{kz}]$$
$$= \Sigma_k m_{kx}, \text{ using (iii) § 70}$$
$$= M_x.$$

$$\therefore\ [M_y M_z] = M_x.$$

These relations found by Dirac's methods were given by Heisenberg and Jordan[1] in the form

$$M_y M_z - M_z M_y = \frac{ih}{2\pi} M_x, \text{ etc.}$$

(v) If M is the resultant angular momentum, then M^2 commutes with M_x, M_y, M_z.

To show this we have

$$\begin{aligned}[M^2 M_x] &= [M_x^2 + M_y^2 + M_z^2, M_x] \\ &= [M_y^2 + M_z^2, M_x], \text{ since } [M_x^2 M_x] = 0, \\ &= [M_y^2 M_x] + [M_z^2 M_x] \\ &= M_y [M_y M_x] + [M_y M_x] M_y + M_z [M_z M_x] \\ &\qquad + [M_z M_x] M_z \\ &= - M_y M_z - M_z M_y + M_z M_y + M_y M_z, \\ &\qquad \text{using (iv) § 71,} \\ &= 0.\end{aligned}$$

Therefore M^2 commutes with M_x; and so with M_y and M_z.

(vi) $[M^2 x] = 2(yM_z - M_y z)$.

For

$$\begin{aligned}[M^2 x] &= [M_x^2 + M_y^2 + M_z^2, x] \\ &= M_x [M_x x] + [M_x x] M_x + M_y [M_y x] + [M_y x] M_y \\ &\qquad + M_z [M_z x] + [M_z x] M_z \\ &= - M_y z - z M_y + M_z y + y M_z, \text{ by (ii) § 71.}\end{aligned}$$

[1] Q.M. II, equation (3), p. 597.

$$\text{But}\quad \left.\begin{array}{l}[M_y z] = \quad x \\ [M_z y] = -x\end{array}\right\} \text{ or } \left\{\begin{array}{l} M_y z - z M_y = \dfrac{ih}{2\pi} x \\ M_z y - y M_z = -\dfrac{ih}{2\pi} x \end{array}\right\}.$$

Using these we have

$$[M^2 x] = 2\,(yM_z - M_y z)$$

or

$$= 2\,(M_z y - z M_y)$$

or

$$= 2\left(M_z y - M_y z + \frac{ih}{2\pi} x\right) = 2\left(y M_z - z M_y - \frac{ih}{2\pi} x\right).$$

(vii) $\frac{1}{2}[M^2[M^2 x]] = -M^2 x - x M^2.$

To show this we have

$$\frac{1}{2}[M^2[M^2 x]] = \left[M^2, M_z y - M_y z + \frac{ih}{2\pi} x\right]$$

$$= M_z[M^2 y] - M_y[M^2 z] + \frac{ih}{2\pi}[M^2 x]$$

(since M^2 commutes with M_y and M_z, § 71 (v)),

$$= 2M_z(zM_x - M_z x) - 2M_y(M_y x - y M_x) + \frac{ih}{2\pi}[M^2 x]$$

$$= 2\,(M_x x + M_y y + M_z z)\,M_x - 2\,(M_x^2 + M_y^2 + M_z^2)\,x + \frac{ih}{2\pi}[M^2 x]$$

since $M_x x = x M_x$ (§ 71 (ii)).

But for an electron (r),

$$m_{xr} x_r + m_{yr} y_r + m_{zr} z_r = 0$$

from the definition of m_x, m_y, m_z (§ 70); hence

$$M_x x + M_y y + M_z z = 0$$

for each electron, using § 71 (i) and proceeding as in § 71 (ii).

$$\therefore\ \frac{1}{2}[M^2[M^2 x]] = -2M^2 x + \frac{ih}{2\pi}[M^2 x]$$

$$= -2M^2 x + [M^2 x - x M^2]$$

$$= -M^2 x - x M^2.$$

72. *Selection and polarisation rules for an atomic system where the forces have symmetry about an axis.*

Suppose the system non-degenerate and take the axis of symmetry as the z axis. Then since the forces have no moment round Oz, $\dot{M}_z = 0$, so that M_z is a diagonal matrix.

Using
$$[M_z z] = 0$$
or
$$M_z z - z M_z = 0,$$
we have
$$\Sigma_k \{M_z(nk)\, z(km) - z(nk)\, M_z(km)\} = 0$$
or
$$M_z(nn)\, z(nm) - z(nm)\, M_z(mm) = 0,$$
since M_z is a diagonal matrix.
$$\therefore\ z(nm)\{M_z(nn) - M_z(mm)\} = 0 \quad \text{.........(1).}$$

So
$$\left.\begin{aligned}[M_z x] &= y\\ [M_z y] &= -x\end{aligned}\right\}\ \text{or}\ \left.\begin{aligned}M_z x - x M_z &= \frac{ih}{2\pi} y\\ M_z y - y M_z &= -\frac{ih}{2\pi} x\end{aligned}\right\},$$
lead to
$$\left.\begin{aligned}x(nm)\{M_z(nn) - M_z(mm)\} &= \frac{ih}{2\pi} y(nm)\\ y(nm)\{M_z(nn) - M_z(mm)\} &= -\frac{ih}{2\pi} x(nm)\end{aligned}\right\} \quad \text{...(2).}$$

Thus from (2), if in a transition $n \to m$, $M_z(nn)$ does not change, $x(nm) = 0$, $y(nm) = 0$, and the light is polarised parallel to the z axis.

But if $M_z(nn)$ does change, then from (1), $z(nm) = 0$.

Also from (2), writing Δ for $\{M_z(mm) - M_z(nn)\}$,
$$\left.\begin{aligned}x(nm)\,\Delta &= -\frac{ih}{2\pi} y(nm)\\ y(nm)\,\Delta &= +\frac{ih}{2\pi} x(nm)\end{aligned}\right\},$$
whence
$$\{x(nm) + iy(nm)\}\left\{\Delta + \frac{h}{2\pi}\right\} = 0,$$
$$\{x(nm) - iy(nm)\}\left\{\Delta - \frac{h}{2\pi}\right\} = 0.$$

Hence we must either have

$$\left.\begin{array}{ll} & \Delta = \dfrac{h}{2\pi}, \text{ and } x(nm) + iy(nm) = 0 \\ \text{or} & \Delta = -\dfrac{h}{2\pi}, \text{ and } x(nm) - iy(nm) = 0 \end{array}\right\} \quad \ldots\ldots(3).$$

Thus if $M_z(nn)$ changes, it can only do so by $\pm \frac{h}{2\pi}$; in each case the light is circularly polarised in the plane xy, and when viewed at right angles to the field appears polarised at right angles to the field.

Thus in general $M_z(nn)$ can only change by 0 or $\pm \frac{h}{2\pi}$.

Therefore $M_z(nn)$ must be of the form $\frac{h}{2\pi}(n_1 + C)$, where $n_1 = \ldots -2, -1, 0, 1, 2, \ldots$ and C is a constant; and also n_1 can only change by 0 or ± 1.

$\Delta n_1 = 0$ corresponds to polarisation parallel to the field and $\Delta n_1 = \pm 1$ to polarisation at right angles to the field.

These results may be summarised as:

$$\left.\begin{array}{l} \begin{matrix} x \\ y \end{matrix}(n_1, n_1) = 0 \\ z(n_1, n_1 \pm 1) = 0 \\ x(n_1, n_1 + 1) + iy(n_1, n_1 + 1) = 0 \\ x(n_1, n_1 - 1) - iy(n_1, n_1 - 1) = 0 \end{array}\right\},$$

the last pair following from equations (3).

73. Since the results of § 72 follow by the use of the formulae

$$[M_z x] = y, \quad [M_z y] = -x, \quad [M_z z] = 0,$$

and since by § 71 (iv)

$$[M_z M_x] = M_y, \quad [M_z M_y] = -M_x, \quad [M_z M_z] = 0,$$

we see that the argument may be repeated with M_x, M_y, M_z taking the place of x, y, z.

Hence we have also for an axial field the results:

$$\left.\begin{aligned}&\begin{matrix}M_x\\M_y\end{matrix}(n_1, n_1) = 0\\&M_z(n_1, n_1 \pm 1) = 0\\&M_x(n_1, n_1+1) + iM_y(n_1, n_1+1) = 0\\&M_x(n_1, n_1-1) - iM_y(n_1, n_1-1) = 0\end{aligned}\right\}.$$

74. *Alternative deduction of the selection and polarisation rules.*

If $M_z = p$ and is the action variable corresponding to the angle variable ϕ (the azimuthal angle about the axis of z), then since

$$[M_z x] = y,\quad [M_z y] = -x,\quad [M_z z] = 0,$$

we have $$[M_z[M_z x]] = [M_z y] = -x.$$

$$\therefore\ [p[px]] = -x$$

So $$\left.\begin{aligned}&[p[py]] = -y\\&[pz] = 0\end{aligned}\right\}\quad\ldots\ldots\ldots\ldots\ldots(1).$$

but

Now expand x as a Fourier series in ϕ, so that

$$x = \Sigma a_n e^{in\phi}.$$

Then on Dirac's theory[1], c_n gives the matrix constituent of x which corresponds to a quantum transition by p of magnitude nh, or to a change of n_1 of magnitude n.

The coefficient c_n contains all the action variables and also all the angle variables except ϕ.

$$\therefore\ [pa_n] = 0 \text{ and } [p, e^{in\phi}] = -ine^{in\phi},$$

the former because p commutes with all the variables in a_n, and the latter by § 82 (v).

$$\therefore\ [px] = \Sigma[p, a_n e^{in\phi}]$$
$$= -i\Sigma n a_n e^{in\phi},$$
$$\therefore\ [p[px]] = -\Sigma n^2 a_n e^{in\phi}.$$

[1] P. A. M. Dirac, Proc. Roy. Soc. A. **110**, § 4, p. 567, **1926**.

But $[p[px]] = -x$, from (1), so that

$$\Sigma a_n e^{in\phi} = \Sigma n^2 a_n e^{in\phi},$$

$$\therefore\ n^2 = 1 \text{ or } a_n = 0.$$

Thus only a_1 and a_{-1} are not zero, and only changes of n_1 by ± 1 lead to amplitudes which are not zero.

Again if $$z = \Sigma c_n e^{in\phi},$$
the condition $[pz] = 0$ of (1) leads to

$$-i\Sigma n c_n e^{in\phi} = 0,$$

so that $c_n = 0$, except when $n = 0$.

Thus only c_0 is not 0, and only non-changes of n_1 lead to amplitudes which do not vanish.

Hence the quantum number n_1 associated with p can only change by 0 or ± 1. If the change is zero there is only a z and the polarisation is parallel to Oz, but if the change is ± 1 there is only an x and y, so that the polarisation is in the plane $z = 0$, and viewed edgewise is seen perpendicular to the field.

CHAPTER XIV

THE THEORY OF THE LANDÉ NUMBERS m, l, j

75. *The nature of the azimuthal quantum number m.*

In the Landé-Pauli theory of the Zeeman effect the angular momentum about the axis of the magnetic field is equal to $\frac{mh}{2\pi}$, where m is the azimuthal quantum number (§ 16). In the notation of § 72 this $m = n_1 + C$ and the selection and polarisation rules for n_1 therefore are those for m, since $\Delta n_1 = \Delta m$. It still remains for the theory to prescribe what the actual values of m can be and not merely to say how m can change.

We therefore consider a free atom acted upon by an axial field and take the axis of symmetry of the field as the z axis. Let the Hamiltonian be $H = H_0 + \lambda H_1 + \dots$, where H_0 is that for the undisturbed atom.

The field does not affect M_z, so that $M_z = M_{z_0}$, where the suffix 0 refers to the undisturbed atom.

Also for the free atom, M_{x_0}, M_{y_0}, M_{z_0} and M_0 are all constant.

M_{x_0}, M_{y_0}, M_{z_0} though constant are not *all* diagonal matrices on account of the relations $[M_{y_0} M_{z_0}] = M_{x_0}$, etc.

or
$$M_{y_0} M_{z_0} - M_{z_0} M_{y_0} = \frac{ih}{2\pi} M_{x_0}, \text{ etc.}$$

For if we take M_{z_0} to be a diagonal matrix, the last equation gives

$$M_{y_0}(nm)\, M_{z_0}(mm) - M_{z_0}(nn)\, M_{y_0}(nm) = \frac{ih}{2\pi} M_{x_0}(nm)$$

or
$$M_{y_0}(nm)\,[M_{z_0}(mm) - M_{z_0}(nn)] = \frac{ih}{2\pi} M_{x_0}(nm) \quad \dots(1).$$

If also M_{y_0} is a diagonal matrix, $M_{y_0}(nm) = 0$. Therefore $M_{x_0}(nm) = 0$.

Also writing $m = n$ in equation (1), $M_{x_0}(nn) = 0$.

Therefore $M_{x_0} = 0$, then from $[M_{x_0} M_{y_0}] = M_{z_0}$ and $[M_{z_0} M_{x_0}] = M_{y_0}$ it follows that $M_{z_0} = 0$ and $M_{y_0} = 0$. Thus if two of the three are diagonal matrices, then all three vanish; thus at most one of the three is a diagonal matrix though all three are constant; this means that the undisturbed system is degenerate.

This degeneracy is removed by the external field, so that the results of § 72 can be used, and by allowing λ to $\rightarrow 0$, properties of the undisturbed atom can be found which relate to the quantum number m representing its potential behaviour in an axial field.

The stationary states of the free atom depend upon a set of quantum numbers (such as l, k, j) whose ensemble is represented by n_2; those of the disturbed atom depend upon n_2 and also upon the n_1 of § 72. The matrix notation (nm) will now mean $(n_1 n_2 m_1 m_2)$ and indicate a transition from the state $n_1 n_2$ to the state $m_1 m_2$ where n_1 has changed to m_1 and at the same time n_2 to m_2.

Since

$$\dot{M}_{x_0} = 0, \; M_{x_0} H_0 - H_0 M_{x_0} = 0 \text{ (§ 52, equation (2))}$$

or
$$M_{x_0}(nm) H_0(mm) - H_0(nn) M_{x_0}(nm) = 0$$

or
$$\nu_0(mn) M_{x_0}(nm) = 0 \quad \text{..................(2).}$$

But since H_0 is independent of n_1,

$$\left.\begin{aligned} \nu_0(n_1 n_2 m_1 n_2) &= 0 \\ \text{and} \qquad \nu_0(n_1 n_2 m_1 m_2) &\neq 0 \end{aligned}\right\}.$$

Therefore from (2),

$$\left.\begin{aligned} M_{x_0}(n_1 n_2 m_1 n_2) &\neq 0 \\ \text{and} \qquad M_{x_0}(n_1 n_2 m_1 m_2) &= 0 \end{aligned}\right\},$$

so that M_{x_0} is a diagonal matrix for n_2.

So for M_{y_0} and M_0^2.

Again since $[M^2 M_z] = 0$ or $M^2 M_z - M_z M^2 = 0$,

$$M^2 (nm) M_z (mm) - M_z (nn) M^2 (nm) = 0.$$

Therefore $M^2 (nm) = 0$, $n \neq m$, so that M^2 is a diagonal matrix for n_1. Hence in the limit for $\lambda \to 0$ this property still holds and M_0^2 is a diagonal matrix for n_1. Thus M_0^2 is a diagonal matrix both for n_1 and n_2.

But M_{z_0} is too, so that $M_{x_0}^2 + M_{y_0}^2$ is also a diagonal matrix for n_1 and n_2 (though M_{x_0}, M_{y_0} separately are only diagonal matrices for n_2).

$$\begin{aligned} \therefore\quad (M_{x_0}^2 + M_{y_0}^2)(n_1 n_2 n_1 n_2) &= \Sigma_{k_1}\Sigma_{k_2}\{M_{x_0}(n_1 n_2 k_1 k_2) M_{x_0}(k_1 k_2 n_1 n_2) \\ &\qquad + M_{y_0}(n_1 n_2 k_1 k_2) M_{y_0}(k_1 k_2 n_1 n_2)\} \\ &= \Sigma_{k_1}\{M_{x_0}(n_1 n_2 k_1 n_2) M_{x_0}(k_1 n_2 n_1 n_2) \\ &\qquad + M_{y_0}(n_1 n_2 k_1 n_2) M_{y_0}(k_1 n_2 n_1 n_2)\}, \end{aligned}$$

(since M_{x_0}, M_{y_0} are each diagonal matrices for n_2, so that for example $M_{x_0}(n_1 n_2 k_1 k_2) = 0$ except for $k_2 = n_2$)

$$= \Sigma_{k_1}\{|M_{x_0}(n_1 n_2 k_1 n_2)|^2 + |M_{y_0}(n_1 n_2 k_1 n_2)|^2\},$$

since the matrices are of Hermite type and therefore

$$M_{x_0}(k_1 n_2 n_1 n_2) = M_{x_0}^*(n_1 n_2 k_1 n_2).$$

Thus the diagonal terms of the matrix $M_{x_0}^2 + M_{y_0}^2$ are always positive, and so therefore are those of the matrix $M_0^2 - M_{z_0}^2$ which is the same quantity.

Therefore since $M_0^2 (n_1 n_2 n_1 n_2)$ does not depend upon n_1, then for a given n_2, $M_0^2 (n_1 n_2 n_1 n_2)$ is fixed. But

$$M_{z_0}^2 (n_1 n_2 n_1 n_2)$$

which is equal to

$$M_z^2 (n_1 n_2 n_1 n_2) \equiv \left(\frac{h}{2\pi}\right)^2 m^2,$$

where $m = n_1 + C$, is steadily increasing with n_1 as we pass along the diagonal of the matrix, so that if

$$M_0^2 (n_1 n_2 n_1 n_2) - M_{z_0}^2 (n_1 n_2 n_1 n_2)$$

is to be positive always, the number of possible values of n_1 and therefore of m is *finite*.

Again using

$$[M_{x_0} M_{y_0}] = M_{z_0} \quad \text{or} \quad M_{x_0} M_{y_0} - M_{y_0} M_{x_0} = \frac{ih}{2\pi} M_{z_0}$$

we have

$$\begin{aligned}\frac{ih}{2\pi} M_{z_0}(n_1 n_2 n_1 n_2) &= \Sigma_{k_1} \Sigma_{k_2} \{M_{x_0}(n_1 n_2 k_1 k_2)\, M_{y_0}(k_1 k_2 n_1 n_2) \\ &\qquad - M_{y_0}(n_1 n_2 k_1 k_2)\, M_{x_0}(k_1 k_2 n_1 n_2)\} \\ &= \Sigma_{k_1} \{M_{x_0}(n_1 n_2 k_1 n_2)\, M_{y_0}(k_1 n_2 n_1 n_2) \\ &\qquad - M_{y_0}(n_1 n_2 k_1 n_2)\, M_{x_0}(k_1 n_2 n_1 n_2)\}.\end{aligned}$$

If, still for a given n_2, the right-hand side of this equation is summed for all possible values of n_1 the result is zero on account of the *finiteness* of n_1 and with it that of k_1. Therefore the left-hand side when summed for all possible values of n_1 is zero.

$$\therefore \quad \Sigma_{n_1} M_{z_0}(n_1 n_2 n_1 n_2) = 0$$

$$\text{or} \quad \frac{h}{2\pi} \Sigma (n_1 + C) = 0 \qquad \text{or} \quad \frac{h}{2\pi} \Sigma m = 0.$$

Thus as the numbers m can only be a series differing by unity and are on account of this last condition symmetrical about zero, they must either be the series

$$\ldots -2, -1, 0, 1, 2, \ldots$$

or the series

$$\ldots -\frac{3}{2}, -\frac{1}{2}, \frac{1}{2}, \frac{3}{2}, \ldots.$$

They are finite in number, the number depending upon n_2, the representative of the quantum numbers of the undisturbed atom. Thus the nature of m is disclosed, and the Landé m numbers for the Zeeman terms accounted for.

The limitation of the m's by the Landé j, involved in the n_2 of this theory (cf. § 13) is also suggested, but it still remains for this theory to show that the limiting value of $|m|$ is actually j, as the Landé theory supposes. This is proved later in § 78.

76. *Quantisation of the resultant angular momentum* M_0.

In the classical theory we transform to axes X, Y, Z where the Z axis is along the resultant angular momentum M (the suffix 0 will be dropped all through but understood) and we take the X axis in the xy plane. Then if l, m, n are the cosines of the Z axis with respect to the axes x, y, z, $l = \frac{M_x}{M}$, etc.

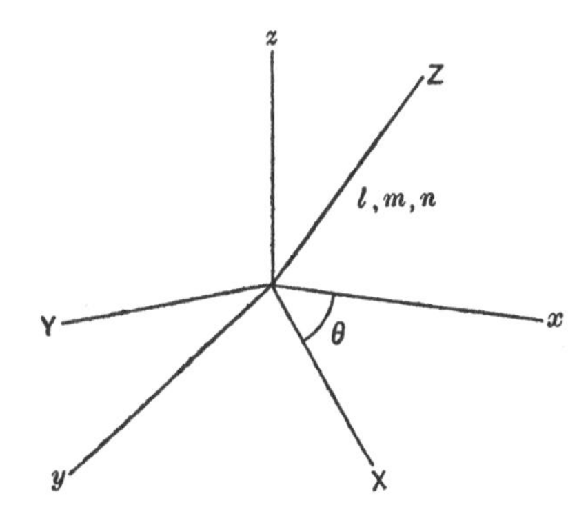

OX has cosines $\cos\theta$, $\sin\theta$, 0 and is perpendicular to OZ;

$$\therefore\ l\cos\theta + m\sin\theta = 0.$$

$$\therefore\ \frac{\cos\theta}{-m} = \frac{\sin\theta}{l} = \frac{1}{\sqrt{l^2+m^2}}.$$

But OY has cosines

$$-n\sin\theta,\ n\cos\theta,\ l\sin\theta - m\cos\theta;$$

hence the scheme:

	x	y	z
X	$\frac{-m}{\sqrt{l^2+m^2}}$	$\frac{l}{\sqrt{l^2+m^2}}$	0
Y	$\frac{-nl}{\sqrt{l^2+m^2}}$	$\frac{-nm}{\sqrt{l^2+m^2}}$	$\sqrt{l^2+m^2}$
Z	l	m	n

$$\therefore\quad X = \frac{ly - mx}{\sqrt{l^2 + m^2}} = \frac{M_x y - M_y x}{\sqrt{M_x^2 + M_y^2}},$$

$$Y = \frac{-nlx - nmy + (l^2 + m^2)z}{\sqrt{l^2 + m^2}}$$

$$= \frac{-M_x M_z x - M_y M_z y + (M_x^2 + M_y^2)z}{M\sqrt{M_x^2 + M_y^2}},$$

$$Z = lx + my + nz = \frac{M_x x + M_y y + M_z z}{M}.$$

These classical forms suggest the following quantum theory transformation:

$$\left.\begin{aligned} X &= yM_x - M_y x \\ Y &= M_x z M_x + M_y z M_y - x M_z M_x - M_y M_z y \\ Z &= xM_x + yM_y + zM_z \end{aligned}\right\}.$$

Then $[M^2 X]$

$$= [M^2, yM_x - M_y x]$$

$$= [M_y x - yM_x, M^2]$$

$$= [M_y M^2]\, x + M_y [xM^2] - [yM^2]\, M_x - y\, [M_x M^2]$$

$$= -M_y [M^2 x] + [M^2 y]\, M_x$$

(since M^2 commutes with M_x and M_y),

$$= -2M_y (M_z y - zM_y) + 2(M_x z - xM_z)\, M_x \quad (\S\, 71\, (\text{vi}))$$

$$= 2(M_x z M_x + M_y z M_y - x M_z M_x - M_y M_z y).$$

$$\therefore\quad [M^2 X] = 2Y.$$

Again $[M^2 Y]$

$$= -[M_x z M_x + M_y z M_y - x M_z M_x - M_y M_z y, M^2]$$

$$= -[(M_x z - xM_z) M_x, M^2] - [M_y (zM_y - M_z y), M^2]$$

$$= -[M_x z - xM_z, M^2]\, M_x - M_y [zM_y - M_z y, M^2]$$

(since M_x, M_y commute with M^2),

$$= \frac{1}{2}[M^2[M^2y]]M_x - \frac{1}{2}M_y[M^2[M^2x]], \qquad \text{§ 71 (vi)}$$

$$= -(M^2y + yM^2)M_x + M_y(M^2x + xM^2), \qquad \text{§ 71 (vii)}$$

$$= M^2(M_yx - yM_x) + (M_yx - yM_x)M^2$$

(since again M_x, M_y commute with M^2),

$$\therefore \; [M^2Y] = -M^2X - XM^2.$$

Also $[M^2Z]$

$$= -[xM_x + yM_y + zM_z, M^2]$$

$$= [M^2x]M_x + [M^2y]M_y + [M^2z]M_z$$

$$= 2\left(yM_z - zM_y - \frac{ih}{2\pi}x\right)M_x + \dots + \dots, \qquad \text{§ 71 (vi)},$$

$$= 2\left\{x\left(M_yM_z - M_zM_y - \frac{ih}{2\pi}M_x\right) + \dots + \dots\right\}$$

$$= 0, \text{ since } [M_yM_z] = M_x, \text{ etc.}$$

$$\therefore \; [M^2Z] = 0.$$

Written in the Heisenberg form[1] these results are:

$$\left.\begin{aligned} M^2X - XM^2 &= \frac{ih}{\pi}Y \\ M^2Y - YM^2 &= -\frac{ih}{2\pi}(M^2X + XM^2) \\ M^2Z - ZM^2 &= 0 \end{aligned}\right\}.$$

If (nm) refers to the quantum number n_2 only, (cf. § 75), then since M^2 is a diagonal matrix for n_2,

$$\left.\begin{aligned} &X(nm)\{M^2(nn) - M^2(mm)\} = \frac{ih}{\pi}Y(nm) \\ &Y(nm)\{M^2(nn) - M^2(mm)\} = \\ &\qquad\qquad -\frac{ih}{2\pi}X(nm)\{M^2(nn) + M^2(mm)\} \\ &Z(nm)\{M^2(nn) - M^2(mm)\} = 0 \end{aligned}\right\} \dots (1).$$

[1] Q.M. II, p. 601, equations (20).

$$\left.\begin{array}{l}\text{If } M^2\,(nn) \text{ does not change } (\Delta n_2 = 0), \text{ then} \\ \qquad\qquad \begin{matrix} X \\ Y \end{matrix}\,(nm) = 0; \\ \text{if } M^2\,(nn) \text{ does change } (\Delta n_2 \neq 0), \text{ then} \\ \qquad\qquad Z\,(nm) = 0 \end{array}\right\} \quad \ldots\ldots(2).$$

In the latter case we eliminate the ratio $X\,(nm) : Y\,(nm)$ and obtain

$$\{M^2\,(nn) - M^2\,(mm)\}^2 - \frac{h^2}{2\pi^2}\{M^2\,(nn) + M^2\,(mm)\} = 0.$$

Write
$$M^2\,(nn) = \left(\frac{h}{2\pi}\right)^2 \left(a_n^2 - \frac{1}{4}\right),$$

so that
$$(a_n^2 - a_m^2)^2 - 2\left(a_n^2 + a_m^2 - \frac{1}{2}\right) = 0.$$

$$\therefore\ [(a_n - a_m)^2 - 1]\,[(a_n + a_m)^2 - 1] = 0.$$

$$\therefore\ a_n = \pm\,(a_m \pm 1).$$

We can without loss of generality take the $+$ sign only outside the last bracket, because only a_n^2 appears in $M^2\,(nn)$.

Therefore $a_n - a_m = \pm\,1$, when $M^2\,(nn)$ changes.

Thus the a's either do not change, or change by $\pm\,1$.

If we write $a_n = j + \frac{1}{2}$, then

$$M^2\,(nn) = \left(\frac{h}{2\pi}\right)^2 j\,(j+1)$$

and the quantum number j can change to $\begin{cases} j+1 \\ j \\ j-1 \end{cases}.$

This j is the quantum number for the angular momentum M and is the Landé j of Zeeman theory.

In the case $j \to j - 1$, $Z\,(j, j-1) = 0$,

but
$$X\,(nm)\,[M^2\,(nn) - M^2\,(mm)] = \frac{ih}{\pi}\,Y\,(nm)$$

becomes

$$X(j, j-1)\left\{\left(\frac{h}{2\pi}\right)^2 j(j+1) - \left(\frac{h}{2\pi}\right)^2 (j-1)j\right\} = \frac{ih}{\pi} Y(j, j-1)$$

or
$$X(j, j-1)\frac{ih}{2\pi} j = - Y(j, j-1) \quad \text{.........(3).}$$

For $j \to j+1$, $Z(j, j+1) = 0$ but

$$X(j, j+1)\left\{\left(\frac{h}{2\pi}\right)^2 j(j+1) - \left(\frac{h}{2\pi}\right)^2 (j+1)(j+2)\right\} = \frac{ih}{\pi} Y(j, j+1)$$

or
$$X(j, j+1)\frac{ih}{2\pi}(j+1) = Y(j, j+1) \quad \text{......(4).}$$

77. *Nature of the quantum number j.*

Since
$$M^2 = M_x^2 + M_y^2 + M_z^2,$$
$$M_x^2 + M_y^2 = M^2 - M_z^2.$$

$$\therefore\ (M_x + iM_y)(M_x - iM_y)$$
$$= M_x^2 + M_y^2 + iM_y M_x - iM_x M_y$$
$$= M_x^2 + M_y^2 + i\left(\frac{ih}{2\pi}\right)[M_y M_x]$$
$$= M^2 - M_z^2 - \frac{h}{2\pi}\{-M_z\}, \qquad \S\ 71\ \text{(iv)}$$
$$= M^2 - M_z^2 + \frac{h}{2\pi} M_z \quad \text{.....................(1).}$$

But from § 73,

$$\left.\begin{aligned}(M_x + iM_y)(m, m+1) = 0\\ (M_x - iM_y)(m, m-1) = 0\end{aligned}\right\} \quad \text{............(2)}$$

where m is used instead of n_1 and now j takes the place of n_2.

Writing $u = M_x + iM_y$, $v = M_x - iM_y$, equation (1) is

$$uv = M^2 - M_z^2 + \frac{h}{2\pi} M_z.$$

We have seen in the course of § 75 that $M_x^2 + M_y^2$ is a diagonal matrix for n_1 and n_2 (here for m, j) and so is M_z, so that uv is a diagonal matrix for m, j.

$$\therefore \quad uv\,(j, m, j, m) = \left(\frac{h}{2\pi}\right)^2 j\,(j+1) - \left(\frac{mh}{2\pi}\right)^2 + \frac{h}{2\pi}\left(\frac{mh}{2\pi}\right)$$

$$= \left(\frac{h}{2\pi}\right)^2 \{j\,(j+1) - m\,(m-1)\}.$$

But $uv\,(j, m, j, m)$ for a given j

$$= uv\,(m, m) \text{ say,}$$

$$= \Sigma_k u\,(mk)\,v\,(km)$$

$$= u\,(m, m-1)\,v\,(m-1, m),$$

since $u\,(mk) = 0$ except when $k = m - 1$,
and $v\,(km) = 0$ except when $k = m - 1$, by (2).

But $$v\,(m-1, m) = v^*\,(m, m-1),$$

since the matrices are of Hermite type,

$$= u\,(m, m-1),$$

since u is the conjugate of v.

$$\therefore \quad uv\,(mm) = \{u\,(m, m-1)\}^2.$$

$$\therefore \quad \{u\,(j, m, j, m-1)\}^2 = \left(\frac{h}{2\pi}\right)^2 \{j\,(j+1) - m\,(m-1)\},$$

$$u\,(j, m, j, m-1) = \frac{h}{2\pi}\sqrt{j\,(j+1) - m\,(m-1)},$$

and since $$v\,(m-1, m) = u\,(m, m-1),$$

$$v\,(j, m-1, j, m) = \frac{h}{2\pi}\sqrt{j\,(j+1) - m\,(m-1)}.$$

$$\left.\begin{aligned} \therefore \quad & (M_x + iM_y)\,(j, m, j, m-1) \\ & \qquad = \frac{h}{2\pi}\sqrt{j\,(j+1) - m\,(m-1)} \\ \text{and} \quad & (M_x - iM_y)\,(j, m-1, j, m) \\ & \qquad = \frac{h}{2\pi}\sqrt{j\,(j+1) - m\,(m-1)} \end{aligned}\right\} \ldots (3).$$

Also $$(M_x - iM_y)(j, m, j, m-1) = 0,$$
and $$(M_x + iM_y)(j, m-1, j, m) = 0.$$

Thus the maximum of m for a given j is fixed by the fact that the transitions $m_{\text{max}} \rightarrow m_{\text{max}} + 1$ must not occur; the right-hand side of (3) thus shows that

$$m_{\text{max}} = j,$$

for if $m = j + 1$, the amplitudes vanish.

Thus on account of the properties of m (§ 75), j *can only be a whole or half integer*. Thus the Landé scheme of numbers for the anomalous Zeeman effect is completely confirmed (cf. § 13).

Quantisation of the angular momentum of the series electron.

This angular momentum is L (cf. § 4) with its quantum number l. In the case of the resultant angular momentum M with its quantum number j, the transition $\Delta j = 0$ means that X, $Y = 0$ and that Z changes, where Z is along M (§ 76). But for the series electron, Z is normal to the plane of the orbit (Z being along L) so that Z cannot change and therefore Δl cannot be zero. Thus whereas Δj can be 0 or $\pm$ 1, Δl can only be $\pm$ 1 (cf. § 4).

CHAPTER XV

INTENSITY FORMULAE FOR THE ZEEMAN EFFECT

78. *Intensity formulae for the Zeeman effect.*

The transformation of § 76 was

$$\left.\begin{aligned} X &= yM_x - M_y x \\ Y &= M_x z M_x + M_y z M_y - x M_z M_x - M_y M_z y \\ Z &= x M_x + y M_y + z M_z \end{aligned}\right\} .$$

We solve for x, y, z in terms of X, Y, Z using the relations $[M_x y] = z$, or $M_x y - y M_x = \epsilon z$, etc., where $\epsilon = ih/2\pi$; in order to obtain the coefficients *behind* the x, y, z.

Thus
$$X = yM_x - (xM_y - \epsilon z).$$
$$\therefore\ yM_x - xM_y = X - \epsilon z.$$

But
$$xM_x + yM_y = Z - zM_z.$$

$$\begin{aligned} \therefore\ (x + iy)(M_x - iM_y) &= Z - zM_z + i(X - \epsilon z) \\ &= Z - z(M_z + i\epsilon) + iX. \end{aligned}$$

$$\therefore\ x + iy = \{Z - z(M_z + i\epsilon) + iX\}(M_x - iM_y)^{-1} \quad (1),$$

and $$x - iy = \{Z - z(M_z - i\epsilon) - iX\}(M_x + iM_y)^{-1} \quad (2).$$

Again

$$\begin{aligned} Y &= (zM_x - \epsilon y) M_x + (zM_y + \epsilon x) M_y - xM_z M_x \\ &\qquad - M_y (yM_z - \epsilon x) \\ &= x(2\epsilon M_y - M_z M_x) - y(\epsilon M_x + M_y M_z) \\ &\qquad + z(M_x^2 + M_y^2 - \epsilon^2), \end{aligned}$$

and using $[M_z M_x] = M_y$, etc., or $M_z M_x - M_x M_z = \epsilon M_y$, etc.,

$$\begin{aligned} Y &= x(\epsilon M_y - M_x M_z) - y(\epsilon M_x + M_y M_z) \\ &\qquad + z(M_x^2 + M_y^2 - \epsilon^2). \end{aligned}$$

Since $X - \epsilon z = yM_x - xM_y$,

$$Y = -xM_xM_z - yM_yM_z - \epsilon(X - \epsilon z) + z(M_x^2 + M_y^2 - \epsilon^2).$$

$$\therefore \quad Y + \epsilon X = -xM_xM_z - yM_yM_z + z(M_x^2 + M_y^2).$$

But $$Z = xM_x + yM_y + zM_z,$$

$$\therefore \quad Y + \epsilon X = -(Z - zM_z)M_z + z(M_x^2 + M_y^2).$$

$$\therefore \quad Y + \epsilon X + ZM_z = zM^2.$$

Therefore multiplying both sides behind by M^{-2},

$$z = (Y + \epsilon X + ZM_z)M^{-2} \quad \text{..............(3).}$$

(1), (2) and (3) determine x, y, z in terms of X, Y, Z.

It can be verified that

$$\left.\begin{aligned} XM_z - M_zX = 0 \\ YM_z - M_zY = 0 \\ ZM_z - M_zZ = 0 \end{aligned}\right\},$$

so that

$$X(nm)\{M_z(mm) - M_z(nn)\} = 0, \text{ or } X(nm) = 0, \ n \neq m.$$

Thus X, and also Y, Z are diagonal matrices with respect to n_1 or with respect to the azimuthal quantum number m.

79. *Consider first transitions where j does not change, or $j \to j$.* Then by (2) § 76

$$X = 0, \ Y = 0,$$

and by (3) § 78 $$z = ZM_zM^{-2};$$

as each factor on the right-hand side is a diagonal matrix with respect to m, we have

$$z(j, m, j, m) = Z(j, m, j, m)\, M_z(j, m, j, m)\, M^{-2}(j, m, j, m)$$

$$= Z(j, m, j, m)\,\frac{mh}{2\pi}\left(\frac{2\pi}{h}\right)^2 \frac{1}{j(j+1)}.$$

Hence, omitting the j from z and Z for brevity,

$$z(mm) = Z(mm)\,.\,\frac{2\pi}{h}\,.\,\frac{m}{j(j+1)} \quad \text{......(1).}$$

Also from (1) § 78

$$x + iy = \{Z - z(M_z + i\epsilon)\}(M_x - iM_y)^{-1} \ldots\ldots (\alpha).$$

We first note that if $v \equiv M_x - iM_y$, then $v^{-1}v = 1$ so that

$$\Sigma_k v^{-1}(mk)\, v(km) = 1.$$

But since $v(km) = 0$ except when $k = m - 1$, this becomes

$$v^{-1}(m, m-1)\, v(m-1, m) = 1.$$

$$\therefore \quad (M_x - iM_y)^{-1}(m, m-1) = v^{-1}(m, m-1)$$

$$= \frac{1}{v(m-1, m)} = \frac{1}{\frac{h}{2\pi}\sqrt{j(j+1) - m(m-1)}}$$

from (3) § 78.

From equation (α), the j not being written in,

$$(x + iy)(m, m-1) = Z(mm)(M_x - iM_y)^{-1}(m, m-1)$$

$$- z(mm)\{M_z(mm) + i\epsilon\}(M_x - iM_y)^{-1}(m, m-1)$$

$$= \frac{Z(mm)}{\frac{h}{2\pi}\sqrt{j(j+1) - m(m-1)}}\left\{1 - \frac{2\pi}{h}\frac{m}{j(j+1)}\left(\frac{mh}{2\pi} - \frac{h}{2\pi}\right)\right\}.$$

$$\therefore \quad (x + iy)(m, m-1)$$

$$= Z(mm)\frac{2\pi}{h}\sqrt{j(j+1) - m(m-1)}/j(j+1) \quad \ldots (2).$$

So $\quad (x - iy)(m-1, m)$

$$= Z(mm)\frac{2\pi}{h}\sqrt{j(j+1) - m(m-1)}/j(j+1) \quad (3).$$

With these too we have always

$$\left.\begin{aligned} (x - iy)(m, m-1) &= 0 \\ \text{and} \qquad (x + iy)(m-1, m) &= 0 \end{aligned}\right\} \quad (\S\ 72).$$

Finally using $[M_x y] = z$ or

$$M_x y - y M_x = \frac{ih}{2\pi} z,$$

it follows that $Z(mm)$ is independent of m.

Thus for the lines

$$j \to j \begin{cases} (x+iy)(j,m,j,m-1) = A\sqrt{j(j+1)-m(m-1)} \\ (x-iy)(j,m-1,j,m) = A\sqrt{j(j+1)-m(m-1)} \\ \qquad z(j,m,j,m) = Am \end{cases}$$

where A is independent of m.

80. *For transitions where* $j \to j-1$, we have $Z=0$ and

$$X(j,j-1)\,\epsilon j = -Y(j,j-1) \qquad (\S\ 76).$$

Also $$z = (Y+\epsilon X)\,M^{-2}.$$

$$\therefore\ z(j,m,j-1,m)$$
$$= Y(j,m,j-1,m)\,M^{-2}(j-1,j-1) + \epsilon X(j,m,j-1,m)\,M^{-2}(j-1,j-1)$$
$$= (-\epsilon j+\epsilon)\,X(j,m,j-1,m)\,.\left(\frac{2\pi}{h}\right)^2 \frac{1}{(j-1)\,j}$$
$$= \frac{1}{\epsilon j}X(j,m,j-1,m).$$

Again $$x+iy = \{iX - z(M_z + i\epsilon)\}(M_x - iM_y)^{-1}.$$

$$\therefore\ (x+iy)(j,m,j-1,m-1) = \{iX(j,m,j-1,m-1) - z(j,m,j-1,m)(M_z(mm)+i\epsilon)\} \times (M_x - iM_y)^{-1}(j-1,m,j-1,m-1)$$
$$= X(j,m,j-1,m)\left\{i - \frac{1}{\epsilon j}\left(\frac{mh}{2\pi} - \frac{h}{2\pi}\right)\right\}\frac{2\pi}{h\sqrt{(j-1)\,j - m(m-1)}}.$$
$$= X(j,m,j-1,m)\left\{\frac{j+m-1}{j}\right\}\frac{2\pi i}{h\sqrt{(j-m)(j+m-1)}}$$

$$\therefore\ (x+iy)(j,m,j-1,m-1) = -\frac{1}{\epsilon j}\sqrt{\frac{j+m-1}{j-m}}\,X(j,m,j-1,m)$$

and so on[1]. The mode of calculation has been sufficiently

[1] Q.M. II, p. 604.

developed, and the results are

$$j \to j-1 \begin{cases} (x+iy)(j,m,j-1,m-1) = -B\sqrt{(j+m)(j+m-1)} \\ (x-iy)(j,m-1,j-1,m) = B\sqrt{(j-m)(j-m+1)} \\ \qquad z\,(j,m,j-1,m) = B\sqrt{j^2-m^2}. \end{cases}$$

So for

$$j \to j+1 \begin{cases} (x+iy)(j,m,j+1,m-1) = -C\sqrt{(j-m+2)(j-m+1)} \\ (x-iy)(j,m-1,j+1,m) = C\sqrt{(j+m+1)(j+m)} \\ \qquad z\,(j,m,j+1,m) = C\sqrt{(j+1)^2-m^2} \end{cases}$$

These formulae agree for large quantum numbers with those found by correspondence principle methods[1]; but for any quantum numbers the corrected formulae of this paragraph are needed, and agree closely with the experimental intensities observed by Ornstein, Burger and Dorgelo[2].

The actual splitting of a multiplet was not found, but has been given in a later paper by Heisenberg and Jordan[3] in which the anomalous Zeeman separation was calculated (chapter XXIV).

[1] S. Goudsmit and R. de L. Kronig, Naturwissensch. **13**, p. 90, **1925**. H. Hönl, Zs. f. Phys. **32**, p. 340, **1925**.

[2] H. B. Dorgelo, Zs. f. Phys. **13**, p. 206, **1923**; **22**, p. 170, **1924**. H. C. Burger and H. B. Dorgelo, Zs. f. Phys. **23**, p. 258, **1924**. L. S. Ornstein and H. C. Burger, Zs. f. Phys. **28**, p. 135; **29**, p. 241, **1924**.

[3] W. Heisenberg and P. Jordan, Zs. f. Phys. **37**, p. 263, **1926**.

CHAPTER XVI

THE CALCULATIONS OF DIRAC AND PAULI FOR HYDROGEN; OPERATOR THEORIES; SCHRÖDINGER'S THEORY

81. *The hydrogen spectrum.*

The foundation of the whole of the quantum theory of spectra is Bohr's theory of the Balmer series; a crucial test of the new mechanics is its ability to yield the Bohr terms Rhc/n^2, where $n = 1, 2, 3, \ldots$. This was first carried out independently by Pauli[1] and Dirac[2], and no little manipulative skill was required to obtain this fundamental result. Pauli required to the full the resources of matrix analysis as developed up to this chapter, and his methods have been sufficiently illustrated for the purpose of this book. Dirac used the action and angle variables of his q-number theory.

The first problem for both was to calculate the Hamiltonian. They assumed that the Hamiltonian has in quantum mechanics the same form as in the classical theory if there occur in it only products of commutable quantities. This is the case generally in *Cartesian* coordinates since for them the Hamiltonian is of the form $H_1(p) + H_2(q)$; for instance, for the Keplerian orbit it is

$$\frac{1}{2m_0}(p_x^2 + p_y^2) - \frac{e^2}{\sqrt{x^2 + y^2}}.$$

In polar coordinates r, θ, the classical Hamiltonian for the Keplerian orbit is

$$\frac{1}{2m_0}\left(p_r^2 + \frac{1}{r^2}p_\theta^2\right) - \frac{e^2}{r},$$

[1] W. Pauli, Zs. f. Phys. **36**, p. 336, Jan. **1926**.
[2] P. A. M. Dirac, Proc. Roy. Soc. A. **110**, p. 561, Jan. **1926**.

and is no longer separable into the parts $H_1(p)$, $H_2(q)$; further there is the difficulty, as Pauli observed, that θ probably cannot be represented by a matrix; it increases without end as the orbital motion proceeds, whereas x, y, r librate between finite limits.

We will now follow Dirac's theory. Starting from the Cartesian form of the Hamiltonian, he transformed it to polar coordinates using q-numbers, and found a polar form quite different from the corresponding form of the classical theory. The details will be given to illustrate Dirac's methods.

82. *Some quantum algebra.*

(i) In the equation for q-numbers $qp - pq = \frac{ih}{2\pi}$, this i is always $+\sqrt{-1}$. Therefore $q^*p^* - p^*q^* = \frac{ih}{2\pi}$, where p^* is the conjugate of p. Therefore $p^*q^* - q^*p^* = -\frac{ih}{2\pi}$. Therefore from one equation we can always find another by writing $-i$ for i and reversing the order of the products.

(ii) $\frac{1}{xy} = \frac{1}{y}.\frac{1}{x}$, as is seen by multiplying by xy in front or behind.

(iii) *The differentiation of* $\frac{1}{x}$.

We have
$$\frac{d}{dt}\left(\frac{1}{x}.x\right) = \frac{d}{dt}(1) = [1H] = 0.$$

$$\therefore\ 0 = \frac{d}{dt}\left(\frac{1}{x}\right).x + \frac{1}{x}\dot{x}.$$

$$\therefore\ \frac{d}{dt}\left(\frac{1}{x}\right)x = -\frac{1}{x}\dot{x},$$

and multiplying by $\frac{1}{x}$ behind,

$$\frac{d}{dt}\left(\frac{1}{x}\right) = -\frac{1}{x}\dot{x}\frac{1}{x}.$$

(iv) The binomial expansion for $(1+x)^n$, if n is a c-number, is the same as in ordinary algebra.

But e^{x+y} is only equal to $e^x e^y$ if x commutes with y.

(v) $[f(q_r), p_r] = \frac{\partial f}{\partial q_r}\frac{\partial p_r}{\partial p_r} = \frac{\partial f}{\partial q_r}$, using the definition of a Poisson bracket.

(vi) If $(aq) = a_1 q_1 + \ldots + a_s q_s$, where the a's are c-numbers, then by (v),

$$[e^{i(aq)}, p_r] = i a_r e^{i(aq)};$$

and since $\quad e^{i(aq)} p_r - p_r e^{i(aq)} = \frac{ih}{2\pi}[e^{i(aq)}, p_r],$

$$e^{i(aq)} p_r = \left(p_r - \frac{a_r h}{2\pi}\right) e^{i(aq)}.$$

From this it follows that

$$e^{i(aq)} f(q_r, p_r) = f\left(q_r, p_r - \frac{a_r h}{2\pi}\right) e^{i(aq)},$$

for if this is true for f_1, f_2 it is true for $f_1 + f_2$ and for $f_1 f_2$; and being true for $f = p_r$ and also for $f = q_r$ (since the q's commute with each other) it is true when f is any power series in the p's and q's.

So $\quad f(q_r, p_r)\, e^{i(aq)} = e^{i(aq)} f\left(q_r, p_r + \frac{a_r h}{2\pi}\right).$

83. *Dirac's calculations for hydrogen.*

The Hamiltonian H

$$= \frac{1}{2m_0}(p_x^2 + p_y^2) - \frac{e^2}{\sqrt{x^2+y^2}},$$

where e, m_0 are c-numbers and the rest q-numbers.

Transform to polars by $x = r\cos\theta$, $y = r\sin\theta$, where $\cos\theta$, $\sin\theta$ are defined in terms of $e^{i\theta}$ by the same relations as in the classical theory. If p_r, p_θ are the momenta conjugate to r, θ, then on the classical theory

$$\left.\begin{aligned} p_r &= p_x\cos\theta + p_y\sin\theta \\ p_\theta &= x p_y - y p_x \end{aligned}\right\}.$$

Dirac therefore for q-numbers tried

$$\left.\begin{aligned} p_r &= \tfrac{1}{2}(p_x \cos\theta + \cos\theta\, p_x) + \tfrac{1}{2}(p_y \sin\theta + \sin\theta\, p_y) \\ p_\theta &= x p_y - y p_x \end{aligned}\right\} \ldots(1).$$

The justification for (1) is that (r, p_r) and (θ, p_θ) should satisfy the conditions for canonical variables, viz.:

$$[rp_r] = 1, \qquad [\theta p_\theta] = 1,$$
$$[r\theta] = 0, \qquad [\theta p_r] = 0, \qquad (\S\ 49\ (\text{iii}))$$
$$[rp_\theta] = 0, \qquad [p_r p_\theta] = 0.$$

By § 54, x, y, r, θ permute with each other.

Also $[rp_x] = [\sqrt{x^2+y^2}, p_x] = \dfrac{\partial}{\partial x}\sqrt{x^2+y^2}$ by § 82 (v)

$$= \frac{x}{\sqrt{x^2+y^2}} = \cos\theta.$$

So $[rp_y] = \sin\theta$.

$$\begin{aligned} \therefore\ [rp_\theta] &= [r, xp_y - yp_x] = [yp_x - xp_y, r] \\ &= y[p_x r] - x[p_y r], \text{ since } x, y \text{ commute with } r, \\ &= -y\cos\theta + x\sin\theta \\ &= -r\sin\theta\cos\theta + r\cos\theta\sin\theta \\ &= 0, \text{ since } \sin\theta \text{ commutes with } \cos\theta. \end{aligned}$$

$$\therefore\ [rp_\theta] = 0.$$

So

$$\begin{aligned} [rp_r] &= \tfrac{1}{2}[r, p_x\cos\theta + \cos\theta\, p_x + p_y\sin\theta + \sin\theta\, p_y] \\ &= -\tfrac{1}{2}[p_x r]\cos\theta - \tfrac{1}{2}\cos\theta[p_x r] - \tfrac{1}{2}[p_y r]\sin\theta \\ &\qquad - \tfrac{1}{2}\sin\theta[p_y r] \\ &= \tfrac{1}{2}\cos^2\theta + \tfrac{1}{2}\cos^2\theta + \tfrac{1}{2}\sin^2\theta + \tfrac{1}{2}\sin^2\theta. \end{aligned}$$

$$\therefore\ [rp_r] = 1.$$

Again

$$\begin{aligned} r[e^{i\theta}, p_\theta] &= [re^{i\theta}, p_\theta], \text{ since } [rp_\theta] = 0, \\ &= [x + iy, xp_y - yp_x] \\ &= [x, xp_y] - [x, yp_x] + i[y, xp_y] - i[y, yp_x] \\ &= y[p_x x] - ix[p_y y] \end{aligned}$$

(since x commutes with x, y, p_y, and y with x, y, p_x),

$$= -y + ix$$
$$= ire^{i\theta}.$$
$$\therefore \quad [e^{i\theta}, p_\theta] = ie^{i\theta}.$$

This agrees with (v) § 82, if θ is canonical to p_θ, or $[\theta p_\theta] = 1$.

So the other conditions $[\theta p_r] = 0$ and $[p_\theta p_r] = 0$ may be verified.

We now solve for p_x, p_y in terms of p_r, p_θ.

Let $u = p_x + ip_y$ and $v = p_x - ip_y$.

Then
$$\left.\begin{aligned} 4p_r &= ue^{-i\theta} + e^{-i\theta}u + ve^{i\theta} + e^{i\theta}v \\ \frac{2i}{r}p_\theta &= e^{-i\theta}u - e^{i\theta}v \end{aligned}\right\} \quad \text{.........(1).}$$

Now $[re^{-i\theta}, u] = [x - iy, p_x + ip_y] = [xp_x] + [yp_y] = 2$, since $[yp_x]$, etc. are zero.

$$\therefore \quad r[e^{-i\theta}, u] + [ru]e^{-i\theta} = 2.$$

But
$$[ru] = [r, p_x + ip_y] = [rp_x] + i[rp_y]$$
$$= \cos\theta + i\sin\theta = e^{i\theta},$$
$$\therefore \quad r[e^{-i\theta}, u] = 2 - 1 = 1,$$
$$[e^{-i\theta}, u] = \frac{1}{r}.$$

So
$$[e^{i\theta}, v] = \frac{1}{r}.$$

$$\left.\begin{aligned} \therefore \quad e^{-i\theta}u - ue^{-i\theta} &= \frac{ih}{2\pi r} \\ e^{i\theta}v - ve^{i\theta} &= \frac{ih}{2\pi r} \end{aligned}\right\} \quad \text{..................(2).}$$

Using (2) in equations (1) in order to get the coefficients of u, v on their *left*, we have

$$2p_r = e^{-i\theta}u + e^{i\theta}v - \frac{ih}{2\pi r},$$
$$\frac{2i}{r}p_\theta = e^{-i\theta}u - e^{i\theta}v.$$

$$\therefore \quad e^{-i\theta} u = p_r + \frac{i}{r} p_\theta + \frac{ih}{4\pi r}.$$

Multiplying both sides by $e^{i\theta}$ on the left, we have

$$\left.\begin{aligned} u &= e^{i\theta}\left(p_r + \frac{i}{r}k_1\right), \text{ where } k_1 = p_\theta + \frac{h}{4\pi} \\ \text{So} \qquad v &= e^{-i\theta}\left(p_r - \frac{i}{r}k_2\right), \text{ where } k_2 = p_\theta - \frac{h}{4\pi} \end{aligned}\right\} \quad \ldots(3).$$

Again $\quad ue^{-i\theta} = e^{-i\theta}u - \dfrac{ih}{2\pi r}$, from (2)

$$= p_r + \frac{i}{r} p_\theta + \frac{ih}{4\pi r} - \frac{ih}{2\pi r}$$

$$= p_r + \frac{i}{r}\left(p_\theta - \frac{h}{4\pi}\right)$$

$$= p_r + \frac{i}{r} k_2.$$

Therefore multiplying through by $e^{i\theta}$ behind,

$$u = \left(p_r + \frac{i}{r}k_2\right)e^{i\theta}, \text{ and so } v = \left(p_r - \frac{i}{r}k_1\right)e^{-i\theta}.$$

Thus we have

$$p_x + ip_y = u = e^{i\theta}\left(p_r + \frac{i}{r}k_1\right) = \left(p_r + \frac{i}{r}k_2\right)e^{i\theta},$$

$$p_x - ip_y = v = e^{-i\theta}\left(p_r - \frac{i}{r}k_2\right) = \left(p_r - \frac{i}{r}k_1\right)e^{-i\theta},$$

where $\quad k_1 = p_\theta + \dfrac{h}{4\pi}$, and $k_2 = p_\theta - \dfrac{h}{4\pi}$.

[Relation (vi) § 82 shows from these that

$$k_2 e^{i\theta} = e^{i\theta} k_1, \quad k_1 e^{-i\theta} = e^{-i\theta} k_2.]$$

Now

$$uv = (p_x + ip_y)(p_x - ip_y) = p_x^2 + p_y^2 + i(p_y p_x - p_x p_y)$$
$$= p_x^2 + p_y^2, \text{ since } p_x \text{ commutes with } p_y.$$

$$\therefore \; p_x^2 + p_y^2 = \left(p_r + \frac{i}{r} k_2\right) e^{i\theta} . e^{-i\theta} \left(p_r - \frac{i}{r} k_2\right)$$

$$= \left(p_r + \frac{i}{r} k_2\right) \left(p_r - \frac{i}{r} k_2\right)$$

$$= p_r^2 + \frac{k_2^2}{r^2} + ik_2 \left(\frac{1}{r} p_r - p_r \frac{1}{r}\right).$$

Now

$$\frac{1}{r} p_r - p_r \frac{1}{r} = \frac{1}{r} (p_r r - r p_r) \frac{1}{r} = \frac{1}{r} \cdot \frac{ih}{2\pi} [p_r r] . \frac{1}{r} = -\frac{ih}{2\pi r^2},$$

since $[r p_r] = 1$.

$$\therefore \; p_x^2 + p_y^2 = p_r^2 + \frac{k_2^2}{r^2} + ik_2 \left(-\frac{ih}{2\pi r^2}\right) = p_r^2 + \frac{k_2}{r^2} \left(k_2 + \frac{h}{2\pi}\right)$$

$$= p_r^2 + \frac{k_1 k_2}{r^2}.$$

$$\therefore \; H = \frac{1}{2m_0} \left(p_r^2 + \frac{k_1 k_2}{r^2}\right) - \frac{e^2}{r}$$

or

$$H = \frac{1}{2m_0} \left\{p_r^2 + \frac{1}{r^2} \left(p_\theta^2 - \frac{h^2}{16\pi^2}\right)\right\} - \frac{e^2}{r} \quad \text{......(4).}$$

If we had originally assumed that the Hamiltonian was the same function of the polar variables as on the classical theory, we should have had instead of (4) that

$$H = \frac{1}{2m_0} \left\{p_r^2 + \frac{1}{r^2} p_\theta^2\right\} - \frac{e^2}{r} \quad \text{............(5).}$$

Dirac then forms the equations of motion

$$\dot{r} = [rH], \;\; \dot{\theta} = [\theta H], \text{ etc.}$$

with each Hamiltonian and finds that the latter leads to a non-degenerate orbit with a precessing apse line, while the former leads to the degenerate elliptic orbit. The latter would lead to a two-fold infinity of energy levels which is contrary to experiment, if the relativity fine structure of the hydrogen spectrum is neglected, as here it is. Hence (4) is the correct Hamiltonian, which agrees with Pauli's form.

In the end, after an intricate calculation, the Bohr formula Rhc/n^2 for the Balmer terms was evolved, with its *integral* quantum numbers 1, 2, 3, ...

Soon after, the problem was solved by the new analysis of Schrödinger which went much further and led to a direct calculation of the *intensities* of the lines (chapter XVII and after).

84. *Difficulties of the matrix calculus.*

In the classical mechanics the Hamiltonian H is transformed to angle and action variables w, J so that H becomes a function of the J's only; it is thus independent of the time which enters only through the w's, and so is a constant. The transformation is effected by the use of a function S which satisfies the Hamilton-Jacobi equation. If this equation can be solved, then the coordinates q can be expressed as functions of the time and H found in terms of the J's. By writing $J_r = n_r h$, where n_r is a quantum number, the energies of the stationary states of the older quantum theory were found[1].

The analogue in the new mechanics is to make H a diagonal matrix so that it is independent of t. $H(p, q)$ is to be transformed to $H(P, Q)$ so that it becomes a diagonal matrix E, whose diagonal terms are the energies of the stationary states. The transformation matrix S, where $S\tilde{S}^* = 1$, gives

$$\left.\begin{aligned} P &= SpS^{-1} \\ Q &= Sq\,S^{-1} \end{aligned}\right\}$$

and
$$E = H(P, Q) = SH(p, q)\,S^{-1}.$$

So far, the difficulties of infinite matrices with their associated infinite number of linear equations with an infinite number of variables (cf. § 68) have proved themselves so great that only cases where the Hamiltonian H

[1] Q.T.A. chap. VII.

is a *perturbation* of a known matrix H_0, has the transformation of H into E by $SHS^{-1} = E$ been effected.

[The solution for hydrogen by Pauli and Dirac was, it is true, directly effected from H but not by any such *general* process; very artful special methods adapted to that particular problem were used.]

Attempts began to be made to bring the theory into connection with some of the more highly developed processes of analysis.

One of the first was due to Lanczos[1] who sought to express the equations of motion and the quantum conditions in the form of integral equations; the theory had the advantage too of being in unison with the usual 'field' ideas of physics.

About the same time too, Born and Wiener[2] also gave a theory in which the properties of a linear operator were used and especially applied to cases of non-periodic phenomena.

85. *Schrödinger's wave mechanics.*

Shortly after, there began to appear a remarkable series of memoirs (January to June 1926) by Erwin Schrödinger[3] in which the solution of a problem in quantum mechanics was made to depend upon the solution of a differential equation, *the wave equation*, in which the energy E of the system appeared. This differential equation only admits of solutions (ψ_n) (which are finite, unique and bounded in the phase space) for certain special values E_n of the energy. These values of E_n are the energies of the

[1] K. LANCZOS, Zs. f. Phys. **35**, p. 812, Dec. **1925**.

[2] M. BORN and N. WIENER, Zs. f. Phys. **36**, p. 174, Jan. **1926**.

[3] E. SCHRÖDINGER, Ann. der Phys. **79**, p. 361, Jan. **1926**; **79**, p. 489, Feb. **1926**; **79**, p. 734, March **1926**; **80**, p. 437, May **1926**; **81**, p. 109, June **1926**. These five papers have been published as the volume 'Abhandlungen zur Wellenmechanik,' by E. SCHRÖDINGER, Barth, Leipzig.

stationary states; the functions ψ_n lead directly to a determination of the Heisenberg matrices for the coordinates by a process of integration (§ 102).

The theory was inspired by the new ideas of Louis de Broglie[1] in which a material particle is associated with a group of waves (chapter XXIII).

[1] L. de Broglie, Ann. de Phys. **10**, p. 22, **1925**. (Thèses, Paris **1924**.) The theory is fully set out in 'Ondes et mouvements,' by L. de Broglie, Gauthier-Villars, Paris **1926**.

CHAPTER XVII

SCHRÖDINGER'S WAVE EQUATION; HIS THEORY OF THE HYDROGEN SPECTRUM

86. *Schrödinger's wave equation.*

The new theory was given in Schrödinger's first paper[1] in much the same casual way as was that of Planck's radiation formula in his earliest papers. Both were arrived at by a process for which no particular justification was given; but the wave equation in the one case and the radiation formula in the other were so striking in their immediate consequences that a real theoretical basis had to be sought for them.

In this first paper the wave equation was found by a variation principle (it was rather like Hamilton's variation principle, but not much more could be said for it); but in the second paper[2] Schrödinger shows it to be a real generalisation of the classical mechanics suggested by the waves of Louis de Broglie (§ 114).

In the former paper Schrödinger remarks that in his theory the 'quantum numbers' appear as naturally as do 'integers' in the theory of a vibrating string, where they are determined by certain boundary conditions to be satisfied by the solution of a differential equation; in quantum mechanics the corresponding differential equation is Schrödinger's wave equation.

In the older quantum theory[3] the Hamilton-Jacobi equation $H\left(q, \frac{\partial S}{\partial q}\right) = E$ is used, where $H(q, p)$ is the Hamiltonian in which $\frac{\partial S}{\partial q}$ is written for p and E is the energy.

[1] E. Schrödinger, Ann. der Phys. **79**, p. 361, **1926**.
[2] E. Schrödinger, Ann. der Phys. **79**, p. 489, **1926**.
[3] Q.T.A. chap. VI.

A solution of it is sought in the form of a *sum* of functions, each of one only of the different q's, i.e.

$$S = f_1(q_1) + f_2(q_2) + \dots + f_s(q_s).$$

Writing $S = K \log \psi$, where K is a constant, so that ψ would be a *product* of functions of the single coordinates, i.e.

$$\psi = \psi_1(q_1)\,\psi_2(q_2) \dots \psi_s(q_s),$$

we obtain the equation

$$H\left(q, \frac{K}{\psi}\frac{\partial \psi}{\partial q}\right) = E.$$

For an electron, using Cartesian coordinates,

$$H = \frac{1}{2m_0}(p_x^2 + p_y^2 + p_z^2) + V,$$

so that the Hamilton-Jacobi equation is

$$\frac{1}{2m_0}\left\{\left(\frac{\partial S}{\partial x}\right)^2 + \left(\frac{\partial S}{\partial y}\right)^2 + \left(\frac{\partial S}{\partial z}\right)^2\right\} + V - E = 0.$$

Putting $S = K \log \psi$, we have

$$\left(\frac{\partial \psi}{\partial x}\right)^2 + \left(\frac{\partial \psi}{\partial y}\right)^2 + \left(\frac{\partial \psi}{\partial z}\right)^2 - \frac{2m_0}{K^2}(E - V)\,\psi^2 = 0.$$

No attempt is made to solve this equation. The expression on the left-hand side ($\equiv \Psi$) was subjected to a variation principle; no particular reason was given for doing so. Schrödinger put

$$\delta \iiint \Psi \, d\tau = 0,$$

the integral being taken over the whole of the q space (here the ordinary x, y, z space) or

$$\delta \iiint \left[\left(\frac{\partial \psi}{\partial x}\right)^2 + \left(\frac{\partial \psi}{\partial y}\right)^2 + \left(\frac{\partial \psi}{\partial z}\right)^2 - \frac{2m_0}{K^2}(E - V)\,\psi^2\right] d\tau = 0.$$

In the usual way, we find

$$\iint dS\,\delta\psi \frac{\partial \psi}{\partial n} - \iiint d\tau\,\delta\psi\left[\frac{\partial^2 \psi}{\partial x^2} + \frac{\partial^2 \psi}{\partial y^2} + \frac{\partial^2 \psi}{\partial z^2} + \frac{2m_0}{K^2}(E - V)\,\psi\right] = 0.$$

Hence

$$\nabla^2\psi + \frac{2m_0}{K^2}(E - V)\psi = 0, \text{ (the wave equation)...(1)},$$

and also
$$\iint dS\,\delta\psi\frac{\partial\psi}{\partial n} = 0$$

over the infinite boundary.

87. *Schrödinger's theory of the hydrogen spectrum.*

The wave equation is

$$\nabla^2\psi + \frac{2m_0}{K^2}(E - V)\psi = 0,$$

and for the hydrogen atom $V = -\frac{e^2}{r}$, so that

$$\nabla^2\psi + \frac{2m_0}{K^2}\left(E + \frac{e^2}{r}\right)\psi = 0 \quad \text{............(1)}.$$

ψ is a product of functions of the separate coordinates (§ 86) and is in just the form needed for the application of harmonic analysis.

Using polar coordinates r, θ, ϕ we write

$$\psi = S_n(\theta, \phi)\,\chi(r),$$

where S_n is a surface spherical harmonic, so that from (1) we have

$$\frac{d^2\chi}{dr^2} + \frac{1}{r}\frac{d\chi}{dr} + \left\{\frac{2m_0 E}{K^2} + \frac{2m_0 e^2}{K^2 r} - \frac{n(n+1)}{r^2}\right\}\chi = 0 \text{...(2)},$$

where $n = 0, 1, 2, 3, \ldots$.

The limitation of n to integers is necessary in order that the dependence of ψ upon θ, ϕ may be unique, for solutions are sought which are continuous, single valued and bounded in the coordinate space (q space). The result found was that such solutions of (2) can be found for *all positive values* of E, but only for a *discrete set of negative values* of E. These values of E are the 'eigenwerte' of the differential equation and the corresponding ψ's are the 'eigenfunctions.'

Thus there is a discrete and a continuous set of eigenfunctions corresponding to certain negative values of E and to all positive values of E. The former gives the Balmer terms of the line spectrum and the latter the continuous spectrum of hydrogen. It is found that for agreement with Bohr's formula, K must be equal to $h/2\pi$.

88. *The solution of the χ equation.*

The equation is

$$\frac{d^2\chi}{dr^2} + \frac{1}{r}\frac{d\chi}{dr} + \left\{\frac{2m_0 E}{K^2} + \frac{2m_0 e^2}{K^2 r} - \frac{n(n+1)}{r^2}\right\}\chi = 0, \text{ (§ 87).}$$

This equation has two singularities: $r = 0$, $r = \infty$. The equation has *in general* no integral which remains finite at both these points, but such an integral exists only for certain values of the coefficients in the differential equation.

Consider first the point $r = 0$. The indicial equation, which determines the behaviour of the integral at this point, is

$$\rho(\rho - 1) + 2\rho - n(n+1) = 0,$$

with roots $\rho = n$ or $-(n+1)$.

The usual solutions in series are one beginning with r^n and another which, owing to the whole number difference between n and $-(n+1)$, contains a logarithm; the latter is therefore of no use in a physical problem. Since the next singular point is at infinity, the former is uniformly convergent and denotes an integral transcendental function.

Write $\chi = r^n U$, so that U is a power series beginning with r^0. The equation becomes

$$\frac{d^2U}{dr^2} + \frac{2(n+1)}{r}\frac{dU}{dr} + \frac{2m_0}{K^2}\left(E + \frac{e^2}{r}\right)U = 0 \quad \text{...(1).}$$

This equation is of the type

$$(a_0 + b_0 x)\frac{d^2y}{dx^2} + (a_1 + b_1 x)\frac{dy}{dx} + (a_2 + b_2 x)y = 0,$$

'the Laplace equation,' with a known theory[1], where

$$\left.\begin{aligned} a_0 &= 0, & b_0 &= 1 \\ a_1 &= 2(n+1), & b_1 &= 0 \\ a_2 &= \frac{2m_0 e^2}{K^2}, & b_2 &= \frac{2m_0}{K^2} E \end{aligned}\right\} \quad \ldots\ldots\ldots\ldots(2).$$

89. *Theory of the Laplace equation.*

The equation is

$$(a_0 + b_0 x)\, y'' + (a_1 + b_1 x)\, y' + (a_2 + b_2 x)\, y = 0.$$

Following the usual method for the solution of a linear equation by a contour integral, we write

$$y = \int_C e^{zx} Z\, dz \quad \ldots\ldots\ldots\ldots\ldots\ldots\ldots(1),$$

where C is a contour, and Z a function of the complex variable z.

Then

$$\int_C e^{zx}\, dz\, Z\, \{(a_0 + b_0 x)\, z^2 + (a_1 + b_1 x)\, z + (a_2 + b_2 x)\} = 0,$$

or

$$\int e^{zx}\, dz\, Z\, (P + Qx) = 0 \quad \ldots\ldots\ldots\ldots\ldots(2),$$

where

$$P = a_0 z^2 + a_1 z + a_2,$$
$$Q = b_0 z^2 + b_1 z + b_2.$$

Now

$$Ze^{zx}(P + Qx) = \frac{d}{dz}(Ze^{zx} Q),$$

provided

$$\frac{d}{dz}(ZQ) = ZP = \frac{P}{Q}(ZQ)$$

or

$$\log(ZQ) = \int_{z_0}^{z} \frac{P}{Q}\, dz,$$

where z_0 is some given z,

or

$$Z = \frac{1}{Q} \exp\left(\int_{z_0}^{z} \frac{P}{Q}\, dz\right) \quad \ldots\ldots\ldots\ldots\ldots\ldots(3).$$

[1] 'Cours d'Analyse,' by E. Goursat, vol. 2, pp. 440–42, Paris, 1905.

With this value of Z, we have from (2)

$$\int_C \frac{d}{dz}(Ze^{zx}Q)\,dz = 0,$$

or

$$(Ze^{zx}Q)_C = 0,$$

or

$$\left\{e^{zx}\exp\left(\int_{z_0}^{z}\frac{P}{Q}\,dz\right)\right\}_C = 0, \text{ using (3)},$$

or

$$\left\{\exp\left(zx + \int_{z_0}^{z}\frac{P}{Q}\,dz\right)\right\}_C = 0 \quad\text{..............(4).}$$

A path C chosen to satisfy this condition (4), will give

$$y = \int_C e^{zx}\frac{1}{Q}\exp\left(\int_{z_0}^{z}\frac{P}{Q}\,dz\right).dz \quad\text{.........(5)}$$

as the solution of the differential equation.

90. *Application of the theory.*

We are concerned with the special case

$$\left.\begin{aligned} P &= a_1 z + a_2 \\ Q &= z^2 + b_2 \end{aligned}\right\} \text{ by § 88 (2),}$$

so that

$$\frac{P}{Q} = \frac{a_1 z + a_2}{z^2 + b_2} = \frac{\alpha_1}{z - c_1} + \frac{\alpha_2}{z - c_2},$$

where

$$c_1 = +\sqrt{-b_2},$$
$$c_2 = -\sqrt{-b_2}.$$

Then

$$\alpha_1 = \frac{a_1 c_1 + a_2}{c_1 - c_2}, \quad \alpha_2 = \frac{a_1 c_2 + a_2}{c_2 - c_1}.$$

$$\therefore \exp\left\{\int_{z_0}^{z}\frac{P}{Q}\,dz\right\} = A\,(z - c_1)^{\alpha_1}(z - c_2)^{\alpha_2},$$

and the constant A may be put equal to 1 as the general solution may always be multiplied by an arbitrary constant.

Therefore from (5),

$$y = \int_C e^{zx}(z - c_1)^{\alpha_1 - 1}(z - c_2)^{\alpha_2 - 1}$$

is the solution, where by (4), C is chosen so that

$$\{e^{zx}(z - c_1)^{\alpha_1}(z - c_2)^{\alpha_2}\}_C = 0.$$

Thus the solution of the χ equation is $\chi = r^n U$, where

$$U = \int_C e^{zr} (z - c_1)^{a_1 - 1} (z - c_2)^{a_2 - 1} dz \quad \text{.........(6)},$$

subject to C being such that

$$\{e^{zr} (z - c_1)^{a_1} (z - c_2)^{a_2}\}_C = 0 \quad \text{...........(7)},$$

where

$$c_1 = + \sqrt{\frac{-2m_0 E}{K^2}}, \qquad c_2 = - \sqrt{\frac{-2m_0 E}{K^2}},$$

$$a_1 = n + 1 + \sqrt{\frac{-m_0 e^4}{2K^2 E}}, \quad a_2 = n + 1 - \sqrt{\frac{-m_0 e^4}{2K^2 E}};$$

or writing
$$l = + \sqrt{\frac{-m_0 e^4}{2K^2 E}},$$

$$a_1 = n + 1 + l, \quad a_2 = n + 1 - l.$$

91. *We have now to consider the behaviour of the solution at the singularity* $r = \infty$.

This is determined by that of the two special independent solutions U_1, U_2 for which C passes once round c_1 or c_2 and begins and ends at infinity at a place such that $e^{zr} \to 0$, i.e. such that the real part of z is $-\infty$. This is needed to satisfy condition (7).

These solutions, for large values of r, are represented asymptotically (in the Poincaré sense) by

$$\left.\begin{aligned} U_1 &\sim e^{c_1 r} r^{-a_1} (-1)^{a_1} (e^{2\pi i a_1} - 1) \Gamma(a_1) (c_1 - c_2)^{a_2 - 1} \\ U_2 &\sim e^{c_2 r} r^{-a_2} (-1)^{a_2} (e^{2\pi i a_2} - 1) \Gamma(a_2) (c_2 - c_1)^{a_1 - 1} \end{aligned}\right\} \text{...(8)},$$

where we retain the first term of the asymptotic series in negative powers of r.

Two cases arise, according as $E \gtrless 0$.

I. $E > 0$.

Here l, c_1, c_2 are pure imaginaries, so that in (8) $e^{c_1 r}$, $e^{c_2 r}$ are finite and U_1 and $U_2 \to 0$ like r^{-n-1}. The same is true of the U we want, as it is linear in U_1 and U_2. Hence $\chi \equiv r^n U \to 0$

like r^{-1}. Thus the χ equation has for all positive values of E continuous, single valued and bounded solutions.

II. $E < 0$.

Two cases of this arise, according as l is or is not an integer.

(i) *l not an integer.*

Then c_1 is real and $+$, c_2 is real and $-$, so that from (8) $U_1 \to \infty$, $U_2 \to 0$ as $r \to \infty$. Thus U can only remain finite at ∞ if it is equal to λU_2, where λ is a numerical factor. But this cannot be so for the following reason. Suppose the path C chosen to be a *closed* path enclosing c_1 and c_2; on account of $\alpha_1 + \alpha_2$ being an integer, condition (7) is satisfied. It is obvious that this contour will give a solution which is finite as $r \to 0$; this is therefore just the U we want. But the closed path can be so deformed as to consist of the two paths giving U_1 and U_2, which leads to the solution $U_1 + e^{2\pi i\alpha_1} U_2$. Thus U cannot be of the form λU_2, but must also contain U_1.

Therefore U cannot remain finite at ∞. Thus there is no solution with the required property of being continuous, single valued and bounded over the whole q-space.

(ii) *l an integer.*

In this case $\sqrt{\dfrac{-m_0 e^4}{2K^2E}} = l$, where $l = 1, 2, 3, \ldots$, so that there is a discrete set of E values given by $E = -\dfrac{m_0 e^4}{2K^2 l^2}$, where $l = 1, 2, 3, \ldots$.

Further $\alpha_1 - 1$, $\alpha_2 - 1$ which are the indices in the integrand of (6) are now integers, respectively equal to $n + l$, $n - l$.

Two sub-cases now arise according as $l \gtrless n$.

(α) $l \leqslant n$.

Then c_1, c_2 are no longer singularities of the integrand in (6), but they can now function as end points of the path C, as condition (7) is then satisfied. A third such point is the negative real ∞. Every path between two of these three points gives a solution. The path from c_1 to c_2 gives a solution regular at $r = 0$, but which $\rightarrow \infty$ as $r \rightarrow \infty$; the two other paths yield only solutions finite as $r \rightarrow \infty$, but infinite at $r = 0$. Thus there is no solution.

(β) $l > n$.

In this case c_1 is a zero, c_2 a pole of at least the first order of the integrand of (6). Two independent integrals are thus given, one by means of the path C passing from $z = -\infty$ to the point c_1 (zero) with avoidance of c_2 (pole), and the other by the residue at the pole. The latter is the U we want.

To find the residue at the pole $z = c_2$, we write, as usual, $z = c_2 + \zeta$ in the integrand of (6), that is in

$$e^{zr} \frac{(z - c_1)^{l+n}}{(z - c_2)^{l-n}},$$

and pick out the coefficient of $1/\zeta$.

The integrand becomes

$$e^{r(c_2+\zeta)} \frac{(2c_2 + \zeta)^{l+n}}{\zeta^{l-n}},$$

since $c_2 = -c_1$, so that we require the coefficient of ζ^{l-n-1} in

$$e^{rc_2} e^{r\zeta} (2c_2)^{l+n} \left(1 + \frac{\zeta}{2c_2}\right)^{l+n},$$

or in

$$e^{rc_2} (2c_2)^{l+n} \Sigma_p \frac{r^p \zeta^p}{p!} \Sigma_q \binom{l+n}{q} \left(\frac{\zeta}{2c_2}\right)^q,$$

or in

$$e^{rc_2} (2c_2)^{l+n} \Sigma_p \Sigma_q \frac{r^p}{p!} \binom{l+n}{q} \left(\frac{1}{2c_2}\right)^q \zeta^{p+q}.$$

Writing $p + q = l - n - 1$, we have as the required residue,

$$e^{rc_2} \Sigma_p \Sigma_q r^p (-2c_1)^{l+n-q} \frac{1}{p!} \binom{l+n}{q} \equiv U,$$

where $\chi = r^n U$.

$$\therefore \quad \chi = e^{-rc_1} \Sigma_{p=0}^{l-n-1} r^{n+p} \frac{(-2c_1)^{2n+1+p}}{p!} \binom{l+n}{l-n-1-p}$$

or writing $rc_1 = x$ or $x = r\sqrt{\dfrac{-2m_0 E}{K^2}}$, we have

$$\chi_{n,l} = e^{-x} x^n \Sigma_{p=0}^{l-n-1} \frac{(-2x)^p}{p!} \binom{l+n}{l-n-1-p} \ldots (9),$$

omitting $(-2c_1)^{n+1}$, a constant factor.

This is the solution which for all real positive values of r remains finite; condition (7) is satisfied on account of the exponential vanishing at ∞.

The eigenfunction ψ, or here $\psi_{n,l}(r, \theta, \phi)$

$$= S_n(\theta, \phi)\, \chi_{n,l}(x),$$

where

$$x = r\sqrt{\frac{-2mE_l}{K^2}}, \quad E_l = -\frac{m_0 e^4}{2K^2 l^2}, \; (l = 1, 2, 3, \ldots),$$

and
$$n = 0, 1, 2, 3, \ldots, l-1.$$

If we write $K = \dfrac{h}{2\pi}$, then

$$E_l = -\frac{2\pi^2 m_0 e^4}{h^2 l^2}, \text{ where } l = 1, 2, 3, \ldots,$$

and these are the well-known energy levels of Bohr which correspond to the hydrogen lines. l is the principal quantum number. Since $n = 0, 1, \ldots, l-1$, it corresponds to an azimuthal quantum number.

If the harmonic $S_n(\theta, \phi)$ is more definitely specified, say as a tesseral T_n^σ, we then have a third quantum number σ naturally introduced.

The eigenwerte E_l for which alone the differential equation has a solution, are the energy levels of the atom; any positive value of E is an eigenwert and these correspond to the continuous spectrum of hydrogen; only discrete negative values of E are eigenwerte and these give the line spectrum.

The eigenfunctions $\chi_{n,l,\sigma}$ enable the Heisenberg matrices corresponding to any quantum variable to be calculated, as will be seen in chapter XX.

Schrödinger regards the $\psi_{n,l,\sigma}$ as associated with some wave phenomenon going on in the atom and as representing some 'vibration form.' Such things as single orbits or the place of an electron in its orbit have no place in his theory; spectral lines are energy emissions due to a change of 'vibration form' and the quantum numbers take their origin in the finiteness and uniqueness of a certain space-function.

CHAPTER XVIII

WAVE MECHANICS

92. *Wave mechanics.*

In the classical dynamics the state of the system whose coordinates are $q_1 \dots q_s$ and momenta $p_1 \dots p_s$ is denoted by a point in $2s$-dimensional space (the q-space); changes in the system are indicated by the passage of the point along some curve in this space, a 'ray,' it may be called. Schrödinger's new concept is that changes in the system should be represented in the q-space not by 'rays' but by 'waves.' He regards the classical mechanics as only an approximation to actual mechanical events, which he says can only be described by wave mechanics; the former in his view bears to the latter just the same relation as geometrical (or 'ray') optics does to physical (or 'wave') optics.

A large scale (macroscopic) mechanical process corresponds to a wave signal in the q-space, which can be regarded as a point in comparison with the geometrical structure of its path. But for small scale (microscopic) processes, such as atomic phenomena are, the rigorous wave theory must be used. The wave equation must replace the usual dynamical equations. The latter are just as unfruitful for the purposes of atom mechanics as geometrical optics alone would be to clear up the fine points of diffraction phenomena.

Hamilton well knew and used the analogy between mechanics and the geometrical optics of the q-space. The Hamilton variation principle $\delta \int L dt = 0$ is Fermat's principle for a wave motion in the q-space; the Hamilton-Jacobi equation expresses Huygens' principle for this wave motion; the new wave mechanics expresses for it the *exact* Kirchhoff analysis of physical optics.

Just as Huygens' principle could deal with the problems of physical optics up to a certain point, so the Hamilton-Jacobi equation could deal with atomic problems up to a point on the older quantum theory (which used the classical mechanics and the correspondence principle).

Just as the exact wave analysis of Kirchhoff was needed to clear up the finer points of physical optics, so the new wave mechanics is required for the *exact* solution of the problems of atom mechanics.

93. *The theory of the wave equation.*

In the classical theory the function S is defined as $\Sigma\int p_k dq_k$, where the p's and q's are the momenta and co-ordinates, so that $p_k = \frac{\partial S}{\partial q_k}$.

Hence $S = \Sigma\int_{t_0}^{t} p_k \dot{q}_k dt = \int_{t_0}^{t} \Sigma \frac{\partial T}{\partial \dot{q}} \dot{q}\, dt = \int_{t_0}^{t} 2\overline{T}\, dt$,

where $\overline{T}$ is the kinetic energy expressed as a function of q, $\dot{q}$.

The action function W is defined as $\int_{t_0}^{t} L\,dt$, where $L = T - V$, V being the potential energy. If E is the constant total energy $\equiv \overline{T} + V$, then

$$W = \int_{t_0}^{t} (2\overline{T} - E)\, dt = S - Et \quad \text{.........(1)},$$

the constant of integration being absorbed in S.

If $H(p, q)$ is the Hamiltonian, the Hamilton-Jacobi[1] equation is

$$H\left(\frac{\partial S}{\partial q}, q\right) = E.$$

[1] Q.T.A. chap. VI.

If $T(p, q)$ is the kinetic energy expressed in terms of p, q, it is a quadratic form in the p's and

$$H(p, q) = T(p, q) + V(q).$$

The Hamilton-Jacobi equation is thus

$$T\left(\frac{\partial S}{\partial q}, q\right) + V - E = 0,$$

or from (1),

$$T\left(\frac{\partial W}{\partial q}, q\right) + V - E = 0 \quad \text{............(2).}$$

In all the applications which follow, $\overline{T}$ is a sum of squares of the $\dot{q}$'s. Thus

$$\overline{T} = \tfrac{1}{2}\Sigma a_k^2 \dot{q}_k^2,$$

where each a is a function of the q's.

But

$$p_k = \frac{\partial \overline{T}}{\partial \dot{q}_k} = a_k^2 \dot{q}_k.$$

$$\therefore \; T(p, q) = \overline{T}(q, \dot{q}) = \tfrac{1}{2}\Sigma p_k^2 / a_k^2 \quad \text{.........(3).}$$

For the geometry in the q-space we suppose the line element ds given by

$$\left.\begin{aligned} & ds^2 = 2\overline{T}(q, \dot{q})\, dt^2 \\ \text{so that here} \quad & ds^2 = \Sigma a_k^2 \dot{q}_k^2\, dt^2 \\ \text{or} \quad & ds^2 = \Sigma a_k^2 dq_k^2 \end{aligned}\right\} \quad \text{..................(4).}$$

$$\therefore \; T\left(\frac{\partial W}{\partial q}, q\right) = \tfrac{1}{2}\Sigma \frac{1}{a_n^2}\left(\frac{\partial W}{\partial q_k}\right)^2, \text{ using (3)}$$

$$= \tfrac{1}{2}\Sigma\left(\frac{\partial W}{a_k \partial q_k}\right)^2$$

$$= \tfrac{1}{2}(\text{grad } W)^2,$$

where grad W is the gradient of W, on account of (4).

Thus equation (2) becomes

$$(\text{grad } W)^2 = 2(E - V) \quad \text{..............(5),}$$

and the geometry is conducted in a non-euclidean space for which the line element (4) depends upon the form of $\overline{T}$.

Now let two surfaces W, $W + dW$ of the family $W =$ constant be drawn at time t. (Since $W = S(q) - Et$, these surfaces change with the time.)

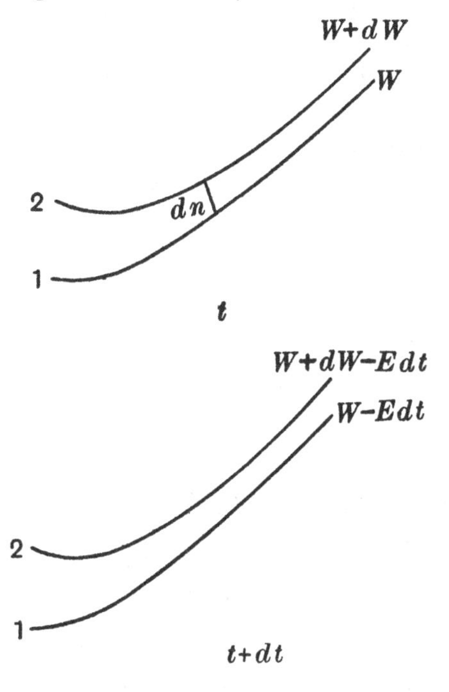

The normal distance (dn) between them is given by

$$dW = \left(\frac{dW}{dn}\right) dn$$

$$= (\text{grad } W)\, dn$$

$$= dn\sqrt{2(E - V)}, \text{ by (5).}$$

$$\therefore \; dn = dW / \sqrt{2(E - V)}.$$

Since $W = S(q) - Et$, the geometrical form of the family of surfaces $W =$ constant is independent of the time, but any individual surface having a value W at time t must have a value $S - E\,(t + dt)$ at time $t + dt$, i.e. the value $W - Edt$.

If we now suppose the surface 1 to move normally to itself in time dt (like a wave front) so as to become the surface 2 and to carry its own W with it, the effect will be the same as above described provided

$$W = W + dW - Edt.$$

(The W of surface 1 in the first figure must be the same as the W of surface 2 in the second figure.)

Therefore $$dW = Edt,$$

or $$\frac{dW}{dn}\,dn = E\,dt,$$

or $$\frac{dn}{dt} = E\bigg/\frac{dW}{dn} = E/\sqrt{2\,(E - V)}, \text{ by (5).}$$

Thus the W-surfaces can be supposed to be wave surfaces in the q-space having a normal velocity

$$v = E/\sqrt{2\,(E - V)} \quad \text{..................(6).}$$

Thus the Hamilton-Jacobi equation is the expression of Huygens' principle in optics, the surface $W + dW$ being constructed from the surface W by drawing spheres of radius $dn \equiv vdt$ and finding their envelope. The refractive index μ is proportional to $\frac{1}{v}$ and varies as $\sqrt{E - V}$; thus the q-space is heterogeneous but isotropic, as μ depends on the coordinates q only.

The Hamilton principle expresses Fermat's principle for the q-space, for the 'least action' form of Hamilton's principle is $\delta\int(\overline{T} - V)\,dt = 0$ subject to

$$\overline{T} + V = E = \text{constant},$$

or $$\delta\int\overline{T}dt = 0,$$

or $$\delta\int ds\,\sqrt{\overline{T}} = 0, \text{ by (4).}$$

But μ varies as $\sqrt{E-V}$, which $= \sqrt{T}$.

Therefore $\delta \int \mu ds = 0$, which is Fermat's principle.

All this shows that the classical mechanics is associated with a geometrical optics in the q-space; clearly a more exact system of mechanics would be one associated with true wave optics in this space, which would give the classical results in cases where the wave length was small compared with the path dimensions.

Schrödinger assumes that a correct extension of the analogy would be to regard the wave system as *sine* waves.

The W function would be $\sin(kW + \alpha)$, where k, α are constants and $W = S - Et$ from (1), so that the W function is $\sin\{k(S - Et) + \alpha\}$.

The frequency ν is therefore $\frac{kE}{2\pi}$; writing $\nu = \frac{E}{h}$, after the manner of de Broglie (§ 114), then $k = \frac{2\pi}{h}$, where h is Planck's constant.

Thus the W function is

$$\sin\left\{\frac{2\pi}{h}(S - Et) + \alpha\right\},$$

where $E = h\nu$. The wave velocity

$$v = E/\sqrt{2(E-V)}, \text{ by (6).}$$

Now for any wave $\Delta\psi = \frac{1}{v^2}\ddot{\psi}$, where Δ is the usual Laplace operator and ψ is the wave function.

$$\therefore \Delta\psi = \frac{2(E-V)}{E^2}\ddot{\psi} \quad \ldots\ldots\ldots\ldots\ldots(7).$$

For a wave of frequency ν, $E = h\nu$, and ψ has a time factor $e^{2\pi i\nu t}$, so that $\ddot{\psi} = -4\pi^2\nu^2\psi$ and

$$\Delta\psi = -\frac{8\pi^2\nu^2(E-V)}{E^2}\psi$$

$$= -\frac{8\pi^2}{h^2}(E-V)\psi.$$

Finally $$\Delta\psi + \frac{8\pi^2}{h^2}(E - V)\psi = 0 \quad\text{...............(8)},$$
which is the wave equation of Schrödinger.

94. *The operator* Δ.

For the coordinate system whose line element ds is given, as above, by

$$ds^2 = a_1{}^2 dq_1{}^2 + a_2{}^2 dq_2{}^2 + \ldots + a_s{}^2 dq_s{}^2,$$

the operator Δ is well known (in potential theory for example) to be

$$\frac{1}{a_1 a_2 \ldots a_s} \Sigma_1{}^s \frac{\partial}{\partial q_k}\left(\frac{a_1 a_2 \ldots a_s}{a_k{}^2} \frac{\partial}{\partial q_k}\right) \quad\text{.........(9)}.$$

Thus for a point mass m_0 in Cartesian coordinates,

$$2\overline{T} = m_0 (\dot{x}^2 + \dot{y}^2 + \dot{z}^2),$$

and therefore

$$ds^2 = 2\overline{T} dt^2 = m_0 (\dot{x}^2 + \dot{y}^2 + \dot{z}^2) dt^2$$
$$= m_0 (dx^2 + dy^2 + dz^2).$$

Thus here a_1, a_2, a_3 are each $\sqrt{m_0}$ and the operator Δ is

$$(m_0)^{-\frac{3}{2}} \left\{\frac{\partial}{\partial x}\left(m_0{}^{\frac{1}{2}} \frac{\partial}{\partial x}\right) + \frac{\partial}{\partial y}\left(m_0{}^{\frac{1}{2}} \frac{\partial}{\partial y}\right) + \frac{\partial}{\partial z}\left(m_0{}^{\frac{1}{2}} \frac{\partial}{\partial z}\right)\right\}$$

or
$$\frac{1}{m_0}\left(\frac{\partial^2}{\partial x^2} + \frac{\partial^2}{\partial y^2} + \frac{\partial^2}{\partial z^2}\right),$$

and the wave equation (8) becomes

$$\frac{\partial^2\psi}{\partial x^2} + \frac{\partial^2\psi}{\partial y^2} + \frac{\partial^2\psi}{\partial z^2} + \frac{8\pi^2 m_0}{h^2}(E - V)\psi = 0,$$

the form used in § 86, with K written equal to $h/2\pi$.

95. *The general form* $(H - E)\psi = 0$, *of the wave equation.*

The wave equation for a coordinate system for which

$$2T = \Sigma_1{}^s a_k{}^2 \dot{q}_k{}^2,$$

and the a's are functions of the q's, is

$$\frac{1}{a_1 a_2 \ldots a_s} \Sigma_1{}^s \frac{\partial}{\partial q_k}\left(\frac{a_1 a_2 \ldots a_s}{a_k{}^2} \frac{\partial\psi}{\partial q_k}\right) + \frac{8\pi^2}{h^2}(E - V)\psi = 0.$$

If we write $-\frac{ih}{2\pi}\frac{\partial}{\partial q_k} = p_k$, this reduces to

$$\frac{1}{a_1 a_2 \ldots a_s}\Sigma_1^s\left(-\frac{2\pi}{ih}\right)p_k\frac{a_1 a_2 \ldots a_s}{a_k^2}\left(-\frac{2\pi}{ih}p_k\psi\right) + \frac{8\pi^2}{h^2}(E-V)\psi = 0,$$

or
$$\left\{\frac{1}{2}\Sigma_1^s\frac{p_k^2}{a_k^2} - E + V\right\}\psi = 0,$$

or
$$\{T(p, q) + V - E\}\psi = 0,$$

using equation (3) of § 93, or

$$\{H(p, q) - E\}\psi = 0,$$

where H is the Hamiltonian.

Thus the general form of Schrödinger's wave equation is

$$\left\{H\left(-\frac{ih}{2\pi}\frac{\partial}{\partial q}, q\right) - E\right\}\psi = 0 \quad \ldots\ldots\ldots(10).$$

Schrödinger shows that even when $2\overline{T}$ is a general quadratic form (and not just a sum of squares), this result holds[1]; the proof requires the tensor calculus.

[1] E. Schrödinger, Ann. der Phys. **79**, equations (31) and (32), p. 748, **1926**.

CHAPTER XIX

SCHRÖDINGER'S THEORY OF THE OSCILLATOR AND ROTATOR; WAVE PACKETS

96. *Schrödinger's theory of the harmonic oscillator*[1].

$T = \frac{1}{2}m\dot{\xi}^2$, $V = \frac{1}{2}m(2\pi\nu_0)^2\xi^2$, where ξ is the displacement.

Write $\xi\sqrt{m} = q$, so that $T = \frac{1}{2}\dot{q}^2$, $V = \frac{1}{2}(2\pi\nu_0)^2 q^2$.

The line element ds is here equal to dq, so that $\Delta = \frac{d^2}{dq^2}$.

The wave equation is

$$\Delta\psi + \frac{8\pi^2}{h^2}(E - V)\psi = 0,$$

or
$$\frac{d^2\psi}{dq^2} + \frac{8\pi^2}{h^2}(E - 2\pi^2\nu_0{}^2 q^2)\psi = 0.$$

$$\therefore \quad \frac{d^2\psi}{dq^2} + (a - bq^2)\psi = 0,$$

where
$$a = \frac{8\pi^2 E}{h^2}, \quad b = \frac{16\pi^4\nu_0{}^2}{h^2}.$$

Write $qb^{\frac{1}{4}} = x$.

Then
$$\frac{d^2\psi}{dx^2} + \left(\frac{a}{\sqrt{b}} - x^2\right)\psi = 0,$$

or
$$\frac{d^2\psi}{dx^2} + (c - x^2)\psi = 0,$$

where
$$c = \frac{a}{\sqrt{b}} = \frac{2E}{h\nu_0} \quad \text{..................(1).}$$

[1] E. Schrödinger, Ann. der Phys. **79**, p. 514, **1926**.

The eigenwerte and eigenfunctions of this equation are well known[1]. The eigenwerte are

$$c = 1, 3, 5, \ldots (2n + 1), \ldots,$$

and the eigenfunctions are the Hermite orthogonal functions $e^{-\frac{x^2}{2}} H_n(x)$, where

$$H_n(x) = (-1)^n e^{x^2} \left(\frac{d}{dx}\right)^n (e^{-x^2})$$

$$= (2x)^n - \frac{n(n-1)}{1!}(2x)^{n-2} + \frac{n(n-1)(n-2)(n-3)}{2!}(2x)^{n-4} - \ldots.$$

From the eigenwert $c_n = (2n + 1)$, we have the corresponding E_n given by $c_n = \frac{2E_n}{h\nu_0}$, from (1).

$$\therefore \quad E_n = h\nu_0 \left(\frac{c_n}{2}\right) = h\nu_0 \left(n + \frac{1}{2}\right),$$

the result of Heisenberg (§ 57).

The eigenfunction

$$\psi_n(\xi) = e^{-\frac{x^2}{2}} H_n(x)\, e^{2\pi i \nu_n t},$$

introducing the time factor, where

$$h\nu_n = E_n = h\nu_0 (n + \tfrac{1}{2}), \text{ or } \nu_n = \nu_0 (n + \tfrac{1}{2}).$$

But $$x = qb^{\frac{1}{4}} = \xi m^{\frac{1}{2}} b^{\frac{1}{4}} = 2\pi\xi \sqrt{\frac{m\nu_0}{h}}.$$

$$\therefore \quad \psi_n(\xi) = e^{-\frac{2\pi^2 m\nu_0}{h}\xi^2} H_n\left(2\pi\xi\sqrt{\frac{m\nu_0}{h}}\right) e^{2\pi i \nu_0 (n+\frac{1}{2})t} \quad \ldots (2).$$

97. *Orthogonal functions.*

A set of functions $\psi_1(q), \psi_2(q), \ldots$, where q stands for all the coordinates $q_1, q_2, \ldots$, form an orthogonal system if

$$\int \psi_n(q)\, \psi_m(q)\, dq = 1, \quad n = m,$$
$$= 0, \quad n \neq m,$$

[1] COURANT and HILBERT, vol. 1, p. 261, 1924.

where dq means $dq_1 dq_2 \ldots$, and the integral is taken over the q-space.

A familiar case is that of a set of zonal harmonics $P_1(\mu), P_2(\mu), \ldots$ which are such that when integrated over the whole range of μ for a sphere, that is from -1 to $+1$,

$$\int P_n P_m d\mu = \frac{2}{2n+1}, \quad n = m,$$
$$= 0, \qquad n \neq m.$$

If each P_n is multiplied by $\sqrt{\frac{2n+1}{2}}$, the above formulae for the new P_n's become

$$\int P_n P_m d\mu = 1, \quad n = m,$$
$$= 0, \quad n \neq m,$$

and the functions are then said to be 'normalised,' the factor $\sqrt{\frac{2n+1}{2}}$ being the 'normalising factor.'

For the Hermite functions $e^{-\frac{x^2}{2}} H_n(x)$ the normalising factor is $\frac{1}{\sqrt{2^n n!}}$, so that if

$$u_n(x) = \frac{1}{\sqrt{2^n n!}} e^{-\frac{x^2}{2}} H_n(x),$$

$$\int_{-\infty}^{\infty} u_n(x)\, u_m(x)\, dx = 1, \quad n = m,$$
$$= 0, \quad n \neq m.$$

98. *Wave packets.*

Schrödinger[1] has used the results of § 96 to illustrate the passage from micro to macro mechanics for the oscillator. He shows that in highly excited states (of large quantum number) a suitably chosen group of eigenfunctions of the type (2), § 96, represents a 'wave packet'

[1] E. Schrödinger, Naturwissensch. **14**, p. 664, **1926**.

which behaves like a point mass of the usual mechanics and oscillates with frequency ν_0 in a rectilinear path.

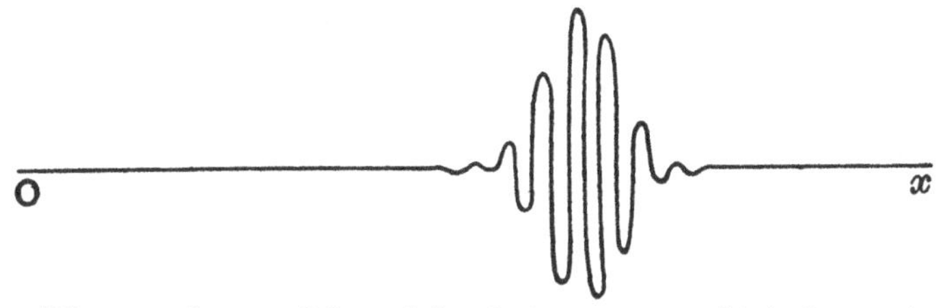

The number and breadth of the waves which form the packet vary with the time, but the width of the packet remains constant, the shape being that of a Gauss error curve. Heisenberg[1] has recently written a critical analysis of the whole question of the passage from micro to macro mechanics; he observes that the Schrödinger result just referred to is *accidentally* true because the frequencies $h\nu_0 (n + \frac{1}{2})$ of the oscillator are separated at equal intervals, as in the classical theory. For a hydrogen atom where the spectral terms heap up to a limit, the wave packet must in Heisenberg's view spread out in course of time over the whole space occupied by the orbit (§ 175), a result quite different from that suggested by Schrödinger's result for the oscillator.

99. *The rigid rotator (molecule).*

The system is two masses m_1, m_2 revolving about their centre of gravity O at constant distances r_1, r_2 from O. The line joining them has polar angles θ, ϕ giving its direction in space.

The kinetic energy

$$\begin{aligned} \overline{T} &= \tfrac{1}{2}m_1 (r_1{}^2\dot{\theta}^2 + r_1{}^2 \sin^2\theta\dot{\phi}^2) + \tfrac{1}{2}m_2 (r_2{}^2\dot{\theta}^2 + r_2{}^2 \sin^2\theta\dot{\phi}^2) \\ &= \tfrac{1}{2}A\,(\dot{\theta}^2 + \sin^2\theta\dot{\phi}^2), \end{aligned}$$

where A is the moment of inertia of the system about a line through O perpendicular to the line m_1m_2.

[1] W. Heisenberg, Zs. f. Phys. **43**, p. 172, § 3, **1927.**

The line element ds of the q-space is given by

$$ds^2 = 2\overline{T}dt^2$$
$$= A\,(d\theta^2 + \sin^2\theta\, d\phi^2),$$

so that $\Delta\psi = \dfrac{1}{A\sin\theta}\left\{\dfrac{\partial}{\partial\theta}\left(\sin\theta\dfrac{\partial\psi}{\partial\theta}\right) + \dfrac{\partial}{\partial\phi}\left(\dfrac{1}{\sin\theta}\dfrac{\partial\psi}{\partial\phi}\right)\right\}$,

using (9), § 94.

Therefore the wave equation is

$$\frac{1}{A\sin\theta}\left\{\frac{\partial}{\partial\theta}\left(\sin\theta\frac{\partial\psi}{\partial\theta}\right) + \frac{1}{\sin\theta}\frac{\partial^2\psi}{\partial\phi^2}\right\} + \frac{8\pi^2}{h^2}(E - V)\,\psi = 0.$$

Here $V = 0$, so that

$$\frac{1}{\sin\theta}\frac{\partial}{\partial\theta}\left(\sin\theta\frac{\partial\psi}{\partial\theta}\right) + \frac{1}{\sin^2\theta}\frac{\partial^2\psi}{\partial\phi^2} + \frac{8\pi^2 AE}{h^2}\psi = 0.$$

Writing $\dfrac{8\pi^2 AE}{h^2} = n\,(n+1)$, we have $\psi_n = S_n\,(\theta, \phi)$, where S_n is a spherical surface harmonic. For ψ_n to be unique and continuous over the whole sphere, n must be $0, 1, 2, \ldots$.

Thus the energy levels are given by

$$\frac{8\pi^2 AE_n}{h^2} = n\,(n+1), \text{ where } n = 0, 1, 2, \ldots,$$

or

$$E_n = \frac{h^2}{8\pi^2 A}\,n\,(n+1), \text{ where } n = 0, 1, 2, \ldots.$$

The earlier quantum theory result[1] was $E_n = \dfrac{h^2 n^2}{8\pi^2 A}$, though the demands of experiment had required half integers to be used for n. This is in agreement with the new formula, for

$$E_n = \frac{h^2}{8\pi^2 A}\left\{\left(n + \frac{1}{2}\right)^2 - \frac{1}{4}\right\},$$

so that energy *differences* are the same as for

$$E_n = \frac{h^2}{8\pi^2 A}\left(n + \frac{1}{2}\right)^2.$$

The eigenfunctions are the harmonics $S_n\,(\theta, \phi)$.

[1] Q.T.A. § 117.
See also F. Reiche and H. Rademacher, Zs. f. Phys. **39**, p. 444, 1926.

CHAPTER XX

THE EVALUATION OF THE HEISENBERG MATRICES BY THE USE OF THE SCHRÖDINGER CALCULUS; DIRAC'S EXTENSION OF HIS THEORY TO RELATIVISTIC MECHANICS

100. *The calculation of the Heisenberg matrices by the use of Schrödinger's eigenfunctions.*

In his third memoir[1] on this subject, Schrödinger begins by pointing out that the Dirac-Heisenberg rules for the p's and q's, viz.:

$$p_r p_s - p_s p_r = 0, \quad q_r q_s - q_s q_r = 0,$$

and

$$\left.\begin{aligned} q_r p_s - p_s q_r &= \frac{ih}{2\pi}, & r &= s \\ &= 0, & r &\neq s \end{aligned}\right\},$$

are equivalent to the usual analysis for *linear differential operators* for the q's if one replaces each p by $-\dfrac{ih}{2\pi}\dfrac{\partial}{\partial q}$.

For $(q_r p_r - p_r q_r)\, u$ becomes

$$= -\frac{ih}{2\pi}\left\{q_r \frac{\partial}{\partial q_r} - \frac{\partial}{\partial q_r} q_r\right\} u$$

$$= -\frac{ih}{2\pi}\left\{q_r \frac{\partial u}{\partial q_r} - \frac{\partial}{\partial q_r}(q_r u)\right\}$$

$$= -\frac{ih}{2\pi}\left\{q_r \frac{\partial u}{\partial q_r} - u - q_r \frac{\partial u}{\partial q_r}\right\}.$$

$$= \frac{ih}{2\pi} u.$$

So $(q_r p_s - p_s q_r)\, u = 0$, $r \neq s$, and so on.

[1] E. Schrödinger, Ann. der Phys. **79**, p. 734, **1926**.

He also points out what has been shown in § 95 that if $H(p, q)$ is the Hamiltonian, the wave equation is

$$\left\{H\left(-\frac{ih}{2\pi}\frac{\partial}{\partial q}, q\right) - E\right\}\psi = 0.$$

Schrödinger using a method of coordinating what he calls a 'well-ordered' function to an operator, where the latter is derived from the former by writing $-\frac{ih}{2\pi}\frac{\partial}{\partial q}$ for each p, shows how to calculate the Heisenberg matrices by the use of the eigenfunctions. But Dirac a few months later gave a theory of this requiring no more than elementary symbolic algebra; this is given in the next article. Thus though the two theories (of Heisenberg and Schrödinger) are so totally different in their inception, they supplement one another and lead to the same results. The great value of Schrödinger's theory is that it brings the determination of the Heisenberg matrices (some of the difficulties of whose calculation we have already encountered) within the scope of the highly developed analysis of differential equation theory, and reduces the calculation to a problem in quadratures only.

101. *Dirac's theory of the derivation of the Heisenberg matrices.*

The Schrödinger wave equation is

$$\left\{H\left(-\frac{ih}{2\pi}\frac{\partial}{\partial q}, q\right) - E\right\}\psi = 0,$$

wherein it is assumed that ψ contains a time factor $e^{-\frac{2\pi i E}{h}t}$, so that

$$\frac{\partial\psi}{\partial t} = -\frac{2\pi i E}{h}\psi \quad\text{or}\quad E\psi = +\frac{ih}{2\pi}\frac{\partial\psi}{\partial t}.$$

Thus the wave equation is

$$(H - E)\psi = 0 \quad \text{......................(1)},$$

where H denotes the operator $H\left(-\frac{ih}{2\pi}\frac{\partial}{\partial q}, q\right)$ and E the operator $\frac{ih}{2\pi}\frac{\partial}{\partial t}$ (cf. § 105 later).

Since (1) is linear in ψ, its general solution is of the form $\Sigma c_n \psi_n$, where the c_n's are arbitrary constants and the ψ_n's are a set of independent solutions (eigenfunctions). [There may be too a continuous set of eigenfunctions $\psi(\alpha)$ depending upon a parameter α and satisfying the differential equation for all values of α in a certain range; or both a continuous and a discrete set may occur together, as for the hydrogen spectrum (§ 87).]

Let a be any constant of integration of the system, i.e. a function of the dynamical variables, such that

$$[a, H - E] = 0, \text{ (cf. § 104)}.$$
$$\therefore\ a(H-E) = (H-E)\,a.$$
$$\therefore\ a(H-E)\,\psi_n = (H-E)\,a\psi_n.$$

But $(H-E)\,\psi_n = 0$. Therefore $(H-E)(a\psi_n) = 0$, so that $a\psi_n$ satisfies the wave equation (1).

Hence it can be expanded in the form

$$a\psi_n = \Sigma_m a_{mn}\psi_m \qquad\qquad (2),$$

where the a_{mn}'s are constants.

We may take the elements a_{mn} to be the elements of the matrix which represents a. For the multiplication law evidently holds, because if $b\psi_n = \Sigma_m b_{mn}\psi_m$, we have

$$\begin{aligned} ab\psi_n &= a\Sigma_m b_{mn}\psi_m \\ &= \Sigma_m b_{mn}(a\psi_m) \\ &= \Sigma_m b_{mn}\Sigma_k a_{km}\psi_k. \end{aligned}$$
$$\therefore\ ab\psi_n = \Sigma_k\Sigma_m a_{km} b_{mn}\psi_k.$$

But
$$ab\psi_n = \Sigma_k (ab)_{kn}\psi_k.$$
$$\therefore\ (ab)_{kn} = \Sigma_m a_{km} b_{mn},$$

the law of matrix multiplication.

As an example of a constant of integration of the dynamical system we may take the value $x(t_0)$ which an arbitrary function x of the variables has at a specified time t_0. The matrix which represents $x(t_0)$ will consist of elements each of which is a function of t_0. Writing t for t_0 we see that an arbitrary function of the dynamical variables $x(t)$ can be represented by a matrix whose elements are functions of t only.

The matrix representation is not unique (cf. chapter XXXI) since any set of independent eigenfunctions ψ_n will do.

To obtain Heisenberg's matrices, the ψ_n's must be chosen so as to make the matrix for E a diagonal matrix.

If $E_1, E_2, \ldots$ are the diagonal terms, then writing E for a in (2) we have

$$E\psi_n = \Sigma_m E_{mn}\psi_m = E_n\psi_n \quad \text{............(3)},$$

since

$$\left.\begin{aligned} E_{mn} &= 0, \quad m \neq n \\ &= E_n, \quad m = n \end{aligned}\right\}.$$

Let x be any function of the dynamical variables which does not contain the time explicitly, and put

$$x\psi_n = \Sigma_m x_{mn}\psi_m,$$

where x_{mn} is a function of t only.

Then

$$\begin{aligned} Ex\psi_n &= \Sigma_m E x_{mn}\psi_m \\ &= \Sigma_m (Ex_{mn} - x_{mn}E)\,\psi_m + \Sigma_m x_{mn}E\psi_m \\ &= \Sigma_m \frac{ih}{2\pi}\dot{x}_{mn}\psi_m + \Sigma_m x_{mn}E_m\psi_m \quad \text{.........(4)}, \end{aligned}$$

using (3) and § 55.

Also since x does not contain t explicitly, $Ex - xE = 0$, and therefore

$$Ex\psi_n = xE\psi_n = xE_n\psi_n = E_n x\psi_n.$$

$$\therefore \; Ex\psi_n = E_n\Sigma_m x_{mn}\psi_m \text{..............(5)}.$$

Equating the coefficients of ψ_m in (4) and (5), we have

$$\frac{ih}{2\pi}\dot{x}_{mn} = x_{mn}(E_n - E_m),$$

so that

$$x_{mn} = a_{mn}e^{\frac{2\pi i}{h}(E_m - E_n)t},$$

where a_{mn} is constant; thus $x_{mn} = a_{mn} \exp 2\pi i \nu (mn) t$, where $h\nu (mn) = E_m - E_n$.

This is the Heisenberg form of matrix element.

Further, $(H - E) \psi_n = 0$, and $E\psi_n = E_n \psi_n$ from (3),

$$\therefore \quad (H - E_n) \psi_n = 0,$$

so that E_n is the eigenwert corresponding to the eigenfunction ψ_n.

Thus in order to express a q-number x as a matrix it is only necessary to expand $x\psi_n$ as a series of terms $\Sigma_m x_{mn} \psi_m$. The coefficient x_{mn} is the mn element of the matrix, which is of the Heisenberg type if the eigenwerte of the problem are taken to be the energy levels of the system.

102. *Calculation of the elements of the matrix for a quantum variable x.*

Suppose the eigenfunctions normalised so that

$$\left.\begin{aligned} \int \bar{\psi}_m \psi_n dq &= 0, \quad m \neq n \\ &= 1, \quad m = n \end{aligned}\right\} \quad \text{..............(1)},$$

where dq means $dq_1 dq_2 \ldots dq_s$, $\bar{\psi}$ is the conjugate of ψ, and the integral is taken over the whole q-space.

We know (§ 101) that if $x\psi_n = \Sigma_m x_{mn} \psi_m$, the coefficient x_{mn} is the (mn) component of the matrix x.

Multiply by $\bar{\psi}_m$ on the two sides and integrate (as for Fourier coefficients) over the q-space. Then

$$\int x\psi_n \bar{\psi}_m dq = \int (\Sigma_m x_{mn} \psi_m) \bar{\psi}_m dq = x_{mn}, \text{ using (1)}.$$

$$\therefore \quad x_{mn} = \int x \bar{\psi}_m \psi_n dq,$$

or

$$x (mn) = \int x \bar{\psi}_m \psi_n dq \text{..................(2)}.$$

The eigenfunctions ψ_n, ψ_m correspond to eigenwerte E_n, E_m which are the energies of the states n, m; the frequency $\nu (mn)$ for a transition $m \rightarrow n$ is $(E_m - E_n)/h$.

From (2) we see that $x (mn) = \bar{x} (nm)$, so that the matrix is of Hermite type.

Thus the eigenfunctions and eigenwerte determine all the amplitudes and frequencies for the quantum variable x. The intensity of the corresponding lines is found from the amplitude by the formula of § 67.

103. *The interpretation of Schrödinger's field scalar* ψ.

It is well known that the electric energy of a system of charges can be expressed as $\frac{1}{2}\Sigma ev$ for the charges, or as a volume integral $\frac{1}{2}\int E^2 dt$. In electrostatics the two expressions are equivalent, but in electrodynamics only the latter has real use. The volume integral is the more general expression for the energy.

On these lines Schrödinger proposes to replace the 'electrons in an atom' by a 'volume density of electricity' which effectively represents the behaviour of the atom when subjected to external influences. The idea of Bohr, Kramers and Slater[1] of supposing the atom in any state to react to external influences in the same way as a virtual radiation field whose frequencies are those associated with all possible transitions from that state to another, is expressed by Schrödinger by using the eigenfunction

$$\psi \equiv \Sigma_n c_n \psi_n (q) \exp 2\pi i E_n t/h,$$

where ψ_n is the eigenfunction corresponding to the stationary state of energy E_n, the c_n's are real constants and the ψ_n's are functions of q, where q stands for all the coordinates $q_1, q_2, \ldots q_s$.

He assumes[2] that the atom reacts to external influences like a *volume density* ρ, where $\rho = \psi\bar{\psi}$, where $\bar{\psi}$ is the conjugate of ψ.

[1] N. Bohr, H. A. Kramers and J. C. Slater, Phil. Mag. **47**, p. 785, **1924**.
[2] E. Schrödinger, Ann. der Phys. **79**, p. 755, equation (36), **1926**.

[The expression used in this equation was $\psi \frac{\partial \bar{\psi}}{\partial t}$; in a later paper in the Ann. der Phys. **80**, p. 476, footnote (2), **1926**, this is corrected to $\psi\bar{\psi}$, the form used above.]

Thus

$$\rho = \Sigma_n c_n \psi_n (q) \exp \{2\pi i E_n t/h\} \Sigma_m c_m \bar{\psi}_m (q) \exp \{- 2\pi i E_m t/h\}$$

$$= \Sigma_{nm} c_n c_m \left\{ \psi_n \bar{\psi}_m \exp \frac{2\pi i t (E_n - E_m)}{h} \right.$$

$$\left. + \psi_m \bar{\psi}_n \exp \frac{2\pi i t (E_m - E_n)}{h} \right\}$$

$$= \Sigma_{nm} c_n c_m [\psi_n \bar{\psi}_m \exp\{2\pi i t \nu (nm)\} + \psi_m \bar{\psi}_n \exp\{-2\pi i t \nu (nm)\}],$$

since $h\nu (nm) = E_n - E_m$.

The electric moment of the atom in the direction of a Cartesian coordinate x is $\int \rho x \, dq$ taken through the q-space, and

$$= \Sigma_{nm} c_n c_m \left[\exp \{2\pi i t \nu (nm)\} \int \bar{\psi}_m \psi_n x \, dq \right.$$

$$\left. + \exp \{- 2\pi i t \nu (nm)\} \int \bar{\psi}_n \psi_m x \, dq \right]$$

$$= \Sigma_{nm} c_n c_m [\exp\{2\pi i t \nu (nm)\} x (mn) + \exp\{-2\pi i t \nu (nm)\} x (nm)],$$

where $x (nm)$ is the nm component of the matrix x, using (2) § 102.

Now $x (nm) = | x (nm) | e^{i\alpha}$, where α is a real phase constant, and

$$x (mn) = \bar{x} (nm) = | x (nm) | e^{-i\alpha}.$$

Hence the electric moment of the atom in the direction of Ox is

$$\Sigma_{nm} c_n c_m | x (nm) | \cos \{2\pi t \nu (nm) - \alpha\},$$

where now the sum is taken only once for each pair of letters n, m.

We have thus a kind of Fourier series for the electric moment in which the frequencies $\nu (nm)$ and the amplitudes $| x (nm) |$ of the Heisenberg analysis occur, which enables the behaviour of the atom in external radiation fields to be calculated.

104. *Dirac's[1] extension of his theory to relativistic mechanics.* Consider a system of s degrees of freedom for which the Hamiltonian involves the time t explicitly.

[1] P. A. M. DIRAC, Proc. Roy. Soc. A. 111, p. 405, 1926.

The principle of relativity demands that t shall be treated as a q-number like the other variables.

On the classical theory it is known that t may be taken as an extra coordinate, with $-E$ as the conjugate momentum.

A Poisson bracket is now defined by

$$[xy] = \Sigma \left(\frac{\partial x}{\partial q}\frac{\partial y}{\partial p} - \frac{\partial x}{\partial p}\frac{\partial y}{\partial q}\right) - \frac{\partial x}{\partial t}\frac{\partial y}{\partial E} + \frac{\partial x}{\partial E}\frac{\partial y}{\partial t} \quad \ldots(1),$$

where now $-E$, t function like a pair p, q.

It is invariant for any contact transformation of the $(2s+2)$ variables. A dynamical system is now determined by an *equation* between these $(2s+2)$ variables (instead of by a *function* of $2s$ variables), viz. the Hamiltonian equation $H - E = 0$, and the equations of motion are

$$\dot{q} = \frac{\partial H}{\partial p} = \frac{\partial}{\partial p}(H - E),$$

$$\dot{t} = 1 = \frac{\partial}{\partial(-E)}(H - E),$$

$$\dot{p} = -\frac{\partial H}{\partial q} = -\frac{\partial}{\partial q}(H - E),$$

$$(-\dot{E}) = (-\dot{H}) = -\Sigma\left(\frac{\partial H}{\partial q}\dot{q} + \frac{\partial H}{\partial p}\dot{p}\right) - \frac{\partial H}{\partial t}$$

$$= -\frac{\partial H}{\partial t}$$

$$= -\frac{\partial}{\partial t}(H - E).$$

From these, if x is any function of the $(2s+2)$ variables,

$$\dot{x} = \Sigma\left(\frac{\partial x}{\partial q}\dot{q} + \frac{\partial x}{\partial p}\dot{p}\right) + \frac{\partial x}{\partial t} + \frac{\partial x}{\partial E}\dot{E}$$

$$= \Sigma\left\{\frac{\partial x}{\partial q}\frac{\partial}{\partial p}(H - E) - \frac{\partial x}{\partial p}\frac{\partial}{\partial q}(H - E) - \frac{\partial x}{\partial t}\frac{\partial(H - E)}{\partial E} + \frac{\partial x}{\partial E}\frac{\partial(H - E)}{\partial t}\right\}$$

$= [x, H - E]$, from the definition (1).

$$\therefore \quad \dot{x} = [x, H - E].$$

These results can be taken over into the quantum theory. We assume $-E$, t to be a new pair of variables (corresponding to a p and q) satisfying the quantum conditions (supplementary to those for the p's and q's):

$$tq - qt = 0, \qquad tp - pt = 0,$$
$$Eq - qE = 0, \qquad Ep - pE = 0,$$
$$-tE + Et = \frac{ih}{2\pi}.$$

The quantum mechanics of moving systems is then developed by Dirac in this paper and applied to work out the theory of the Compton effect. The comparison of Dirac's results with experiment is given later (§ 118).

105. *Schrödinger's wave equation in relativistic form.*

For a particle of 'rest' mass m_0 moving in free space with velocity v, the energy $E = \frac{m_0 c^2}{\sqrt{1-\beta^2}}$, where $\beta = \frac{v}{c}$, and c is the velocity of light (§ 114).

Also[1] $p_x = \frac{m_0 \dot{x}}{\sqrt{1-\beta^2}}$, etc.

$$\therefore \quad \Sigma p_x^2 = \frac{m_0^2 v^2}{1-\beta^2} = \frac{m_0^2 c^2 \beta^2}{1-\beta^2} = m_0^2 c^2 \left[\frac{1}{1-\beta^2} - 1\right].$$

$$\therefore \quad \Sigma p_x^2 = -m_0^2 c^2 + \frac{E^2}{c^2}.$$

$$\therefore \quad \Sigma p_x^2 + m_0^2 c^2 - \frac{E^2}{c^2} = 0.$$

Writing $p_x = -\frac{ih}{2\pi}\frac{\partial}{\partial x}$, etc., $(-E) = -\frac{ih}{2\pi}\frac{\partial}{\partial t}$, in the expression on the left-hand side and allowing it to operate on ψ we have the Schrödinger wave equation

$$\left(\frac{\partial^2}{\partial x^2} + \frac{\partial^2}{\partial y^2} + \frac{\partial^2}{\partial z^2} - \frac{4\pi^2}{h^2} m_0^2 c^2 - \frac{1}{c^2}\frac{\partial^2}{\partial t^2}\right)\psi = 0.$$

[1] Q.T.A. § 36.

CHAPTER XXI

PERTURBATION THEORY IN WAVE MECHANICS; THE INTENSITIES IN THE STARK EFFECT

106. *Perturbation theory in wave mechanics.*

The problem is to determine the eigenfunctions for a slightly perturbed system when those of the original system are known.

The method used[1] is suggested by that of Lord Rayleigh[2] for finding the vibrations of a string of slightly varying density, where the new normal functions are found when the old ones are known for a uniform string.

Schrödinger's equation for the undisturbed system is

$$\left\{H\left(-\frac{ih}{2\pi}\frac{\partial}{\partial q}, q\right) - E\right\}\psi = 0$$

or

$$(H - E)\,\psi = 0 \quad \text{.................}(1),$$

where H denotes the above operator.

The eigenfunctions and eigenwerte for this equation are supposed known; let them be

$$\psi_s\,(q),\ E_s,\ \text{where } s = 1, 2, 3, \ldots.$$

We also suppose the eigenfunctions normalised so that

$$\left.\begin{aligned}\int \bar{\psi}_s \psi_k dq &= 0, \quad s \neq k \\ &= 1, \quad s = k\end{aligned}\right\} \quad \text{...............}(2).$$

We have now to find how ψ_s and E_s are affected by a small perturbation which adds a small term $\lambda A \psi$ to the left-hand side of (1), where λ is a small constant and A is a function of the q's.

[1] E. Schrödinger, Ann. der Phys. **80**, p. 437, **1926**.

[2] Rayleigh, 'Theory of Sound,' vol. 1, 2nd edition, pp. 115–118, London, **1894**.

The wave equation for the perturbed system is

$$(H + \lambda A - E)\,\psi = 0 \quad \text{..............(3)}$$

and let the new eigenfunctions and eigenwerte be ψ_s', E_s' $(s = 1, 2, 3, \ldots)$, where

$$E_s' = E_s + \lambda\epsilon_s + \ldots,$$
$$\psi_s' = \psi_s + \lambda v_s + \ldots.$$

Substituting in (3), which for a given s is

$$(H + \lambda A - E_s')\,\psi_s' = 0,$$

and neglecting λ^2, as only a first approximation is proposed, we have

$$\{H - E_s + \lambda\,(A - \epsilon_s)\}\,(\psi_s + \lambda v_s) = 0$$

or $$(H - E_s)\,\psi_s + \lambda\,\{(H - E_s)\,v_s + (A - \epsilon_s)\,\psi_s\} = 0.$$

But $(H - E_s)\,\psi_s = 0$.

$$\therefore\ (H - E_s)\,v_s = -\,(A - \epsilon_s)\,\psi_s \text{...........(4)}.$$

This equation has only a solution[1] (satisfying the usual conditions) if the right-hand side is orthogonal to the eigenfunction of the left-hand side equated to zero, that is, orthogonal to ψ_s.

[For the string problem referred to above, the equation corresponding to equation (4) would be

$$\ddot{\phi}_s + p_s^2\phi_s = F\cos qt.$$

The eigenfunction of the left-hand side equated to zero is $\cos p_s t$ and if the disturbance ϕ_s is to be finite there must be no resonance, or $p_s \neq q$, i.e.

$$\int_0^\tau \cos p_s t \cos qt = 0,$$

where τ is a period. This corresponds exactly to the orthogonal condition above.]

Hence here for a finite solution of (4),

$$\int (A - \epsilon_s)\,\psi_s\bar{\psi}_s dq = 0.$$

[1] COURANT and HILBERT, chap. v, p. 277.

$$\therefore \quad \epsilon_s = \int A\psi_s\bar{\psi}_s\, dq \quad \ldots\ldots\ldots\ldots\ldots\ldots (5),$$

since $\int \psi_s \bar{\psi}_s dq = 1$, by (2).

[This result is the parallel of the classical theory, that the energy perturbation $\lambda\epsilon_s$ to a first approximation is equal to the perturbation function λA averaged over the undisturbed path.]

To find the perturbation λv_s of the eigenfunction, we have to solve (4) which is

$$(H - E_s)\, v_s = -\,(A - \epsilon_s)\, \psi_s.$$

Let v_s and $(A - \epsilon_s)\, \psi_s$ when expanded in a series of the eigenfunctions ψ_k be

$$\left.\begin{aligned} v_s &= \Sigma_k v_{ks} \psi_k \\ (A - \epsilon_s)\, \psi_s &= \Sigma_k a_{ks} \psi_k \end{aligned}\right\} \ldots\ldots\ldots\ldots\ldots\ldots (6).$$

From the latter, multiplying all through by $\bar{\psi}_k$ and integrating over the q-space, we have on account of the normal property of the ψ's (equation 2),

$$\int (A - \epsilon_s)\, \bar{\psi}_k \psi_s\, dq = a_{ks}$$

or

$$\int A\bar{\psi}_k\psi_s dq - \epsilon_s \int \bar{\psi}_k \psi_s dq = a_{ks}.$$

$$\left.\begin{aligned} \therefore \quad a_{ks} &= \int A\bar{\psi}_k\psi_s dq \quad (s \neq k) \\ \text{or } &= 0 \qquad\qquad\qquad (s = k) \end{aligned}\right\} \ldots\ldots\ldots (7),$$

on account of (5).

Substituting from (6) into (4), we have

$$\Sigma_k v_{ks}\, (H - E_s)\, \psi_k = -\, \Sigma_k a_{ks} \psi_k,$$

and since $(H - E_k)\, \psi_k = 0$,

$$\Sigma_k v_{ks}\, (E_k - E_s)\, \psi_k = -\, \Sigma_k a_{ks} \psi_k.$$

$$\therefore\ v_{ks} = \frac{a_{ks}}{E_s - E_k} = \frac{\int A\bar{\psi}_k\psi_s dq}{E_s - E_k} \quad (s \neq k) \ \ldots\ldots (8).$$

Normalising ψ_s', we have

$$\int (\psi_s + \lambda v_s)(\bar{\psi}_s + \lambda \bar{v}_s)\, dq = 1,$$

or
$$\int \psi_s \bar{\psi}_s dq + \lambda \int (v_s \bar{\psi}_s + \bar{v}_s \psi_s)\, dq = 1,$$

or
$$\int (v_s \bar{\psi}_s + \bar{v}_s \psi_s)\, dq = 0,$$

or
$$\int \{\bar{\psi}_s \Sigma v_{ks} \psi_k + \psi_s \Sigma \bar{v}_{ks} \bar{\psi}_k\}\, dq = 0,$$

or $\qquad v_{ss} + \bar{v}_{ss} = 0$, using (2). $\quad \therefore v_{ss} = 0.$

Thus for the perturbed system the eigenfunctions are

$$\psi_s + \lambda \Sigma_k' \psi_k \frac{\int A\, \bar{\psi}_k \psi_s dq}{E_s - E_k},$$

and the eigenwerte are

$$E_s + \lambda \int A \psi_s \bar{\psi}_s dq,$$

where the dash above the Σ means that the term $k = s$ is omitted.

107. *Degenerate systems.*

Schrödinger[1] then extends this theory to the case of several variables and also gives the perturbation theory of a degenerate system. Such a system would have, corresponding to an eigenwert E_s, several eigenfunctions

$$\psi_{s_1}, \psi_{s_2}, \ldots \psi_{s_n}.$$

The disturbed system has then new eigenwerte

$$E_s + \epsilon_{s_1}, \ldots E_s + \epsilon_{s_n},$$

with corresponding eigenfunctions

$$\psi_{s_1} + v_{s_1}, \ldots \psi_{s_n} + v_{s_n}$$

and the energy level E_s splits up into n new levels.

108. *Schrödinger's calculation of the intensities in the Stark effect.*

In this fourth memoir Schrödinger applies the perturbation theory to work out the Stark effect of an electric

[1] E. Schrödinger, l.c. § 106.

field on the hydrogen lines, and calculates the intensities of the components of the lines H_α, H_β, H_γ, H_δ.

He uses the Epstein coordinates[1]

$$x = \sqrt{\lambda\mu}\cos\phi, \quad y = \sqrt{\lambda\mu}\sin\phi, \quad z = \frac{1}{2}(\lambda - \mu).$$

The wave equation

$$\Delta\psi + \frac{8\pi^2}{h^2}\left(E + \frac{e^2}{r} - eFz\right)\psi = 0,$$

where F is the electric field, parallel to Oz, transforms into

$$\frac{\partial}{\partial\lambda}\left(\lambda\frac{\partial\psi}{\partial\lambda}\right) + \frac{\partial}{\partial\mu}\left(\mu\frac{\partial\psi}{\partial\mu}\right) + \frac{1}{4}\left(\frac{1}{\lambda} + \frac{1}{\mu}\right)\frac{\partial^2\psi}{\partial\phi^2}$$
$$+ \frac{2\pi^2 m}{h^2}\left[E(\lambda + \mu) + 2e^2 - \frac{1}{2}cF(\lambda^2 - \mu^2)\right]\psi = 0.$$

The perturbation theory leads to eigenwerte

$$E = -\frac{2\pi^2 me^4}{h^2 l^2} - \frac{3}{8}\frac{h^2 Fl(n_1 - n_2)}{\pi m^2 e},$$

where $l = n_1 + n_2 + n + 1$.

The corresponding eigenfunctions (to a 'zero' approximation) are

$$\psi_{n n_1 n_2} = (\lambda\mu)^{\frac{n}{2}} e^{-\frac{\lambda+\mu}{2la}} L^n_{n+n_1}\left(\frac{\lambda}{la}\right) L^n_{n+n_2}\left(\frac{\mu}{la}\right) \begin{matrix}\sin \\ \cos\end{matrix}\; n\phi,$$

where $a = \dfrac{h^2}{4\pi^2 me^2}$ = radius of Bohr's 'ground' orbit for hydrogen[2] and $L^n_{n+\sigma}(x)$ is the nth differential coefficient of the $(n + \sigma)$th Laguerre polynomial[3].

The numbers n_1, n_2, $n + 1$ are the n_1, n_2, n_3 of Epstein.

The matrix for a coordinate x is then given by

$$x(nn_1n_2, mm_1m_2) = \int dq\,\bar{\psi}_{nn_1n_2}\psi_{mm_1m_2}x \quad (\S\ 102\ (2))$$

taken through the coordinate space λ, μ, ϕ.

[1] Q.T.A. chap. IX. [2] Q.T.A. p. 23.
[3] COURANT and HILBERT, chap. II, p. 78.

Also $dq = \frac{\partial (xyz)}{\partial (\lambda\mu\phi)} d\lambda d\mu d\phi = \frac{1}{2}(\lambda + \mu)\, d\lambda d\mu d\phi,$

and the ψ's are supposed normalised before use in the equation.

The corresponding intensity is then given by

$$I(nn_1n_2, mm_1m_2) = \frac{[2\pi\nu(nn_1n_2, mm_1m_2)]^4}{3c^3} \times |x(nn_1n_2, mm_1m_2)|^2 \quad (\S\ 67)$$

for a line due to a transition from the state n, n_1, n_2 to the state m, m_1, m_2.

Such is the outline of Schrödinger's work, from which by laborious calculation were found the numbers given below (§ 110); modern atomic theory is rapidly becoming dependent, like astronomy, upon the elaborate system of computation of the observatory.

109. *The selection and polarisation rules* for the lines, in the form given by Epstein, are seen at once from the eigenfunctions.

For $x(nn_1n_2, mm_1m_2) = \iiint dq\, \bar{\psi}_{nn_1n_2} \psi_{mm_1m_2}\, x$

$$= \iiint \frac{1}{2}(\lambda + \mu)\, d\lambda d\mu d\phi\, \bar{\psi}_{nn_1n_2} \psi_{mm_1m_2} (\sqrt{\lambda\mu} \cos\phi).$$

The integration with respect to ϕ on the right-hand side is seen, by looking at the expressions for $\psi_{nn_1n_2}$ (§ 108), and using the lower value, to be

$$\int_0^{2\pi} \cos\phi \cos n\phi \cos m\phi\, d\phi,$$

and this is obviously zero unless n and m differ by 1, or $n - m = \pm 1$. Thus the x intensity vanishes (and so the y intensity too) except when $\Delta n = \pm 1$. Thus there is circular polarisation in the xy plane when $\Delta n = \pm 1$.

So for $z\,(nn_1n_2,\, mm_1m_2)$, we have z for x on the right-hand side and as $z = \frac{1}{2}(\lambda - \mu)$, the ϕ integral is just

$$\int_0^{2\pi} \cos n\phi \cos m\phi \, d\phi$$

and vanishes except when $m = n$ or $\Delta n = 0$. Thus there is polarisation parallel to Oz when $\Delta n = 0$.

In the course of the work which leads to the eigen-function $\psi_{nn_1n_2}$ it appears that n must be $0, 1, 2 \ldots$, so that Epstein's equatorial quantum number n_3, being $n + 1$, can only be $1, 2, 3 \ldots$.

Thus the value $n_3 = 0$, excluded by Epstein[1] by supposing the non-existence of paths which would lead to collision with the nucleus, never appears in this theory (nor does it on Heisenberg's theory as applied by Pauli[2] to find the Stark effect separations), as there is no eigen-function. These results correspond to Bohr's formula

$$E = -\frac{2\pi^2 me^4}{h^2 l^2} - \frac{3}{8}\frac{h^2 Flk}{\pi^2 me},$$

so that Bohr's $k = n_1 - n_2$. Also $l = n_1 + n_2 + n + 1$.

$$\therefore \; l + k = 2n_1 + n + 1.$$

$$\therefore \; \Delta\,(l + k) = 2\Delta n_1 + \Delta n.$$

Therefore if $\Delta n = \pm 1$, $\Delta\,(l + k)$ is odd and the polarisation is in the xy plane; if $\Delta n = 0$, $\Delta\,(l + k)$ is even and the polarisation is parallel to z. This is the form in which Bohr states the selection principle[3].

110. *Schrödinger's numerical results.*

The following is a table of the results for the Stark components for H_α. The numbers for the observed intensities are relative, as are those for the theoretical ones, with different scales for the two sets.

[1] Q.T.A. § 73.
[2] W. Pauli, Zs. f. Phys. **36**, p. 358, **1926**.
[3] Q.T.A. § 136.

Polarisation	*No. of Compt.*	*Exp. Intensity*	*Theor. Intensity*	
∥ to field	2	1	729	Total 4715
	3	1·1	2304	
	4	1·2	1681	
	8	not observed	1	
⊥ to field	0	1·3	2745	Total 4715
	1	1	1936	
	5	not observed	16	
	6	not observed	18	

These are shown in the figure below:

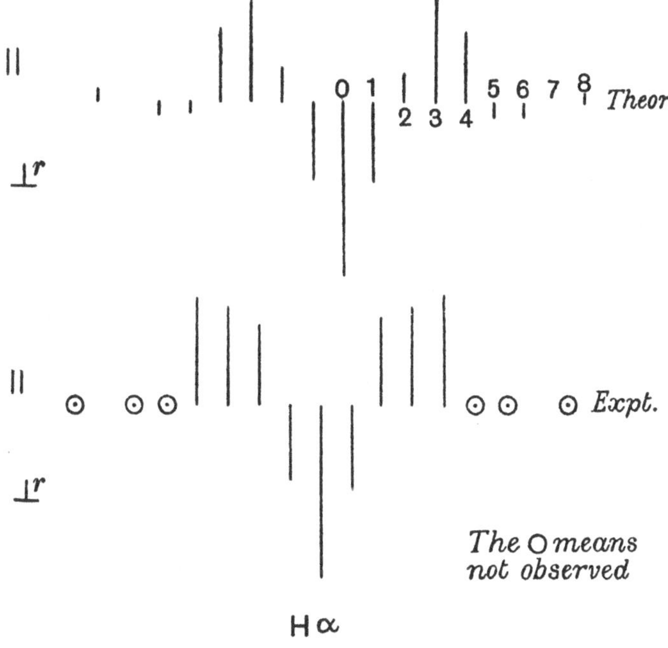

Similar results were given for H_β, H_γ and H_δ.

Schrödinger's figures show that the agreement with experiment is tolerably good, and better than results deduced by the correspondence principle in the early days of exploration of spectral theory by Bohr and Kramers[1]; certain serious contradictions of that theory with experiment are removed on this theory.

Schrödinger's theoretical intensities also fulfil a fundamental condition found by experiment[2] that the sum of the intensities of the parallel components is equal to that of the perpendicular components (see the table above). The 'total intensities' of the four hydrogen lines considered (each 'total intensity' being the sum of the intensities of the Stark components) were found to be in the ratio

$$H_\alpha : H_\beta : H_\gamma : H_\delta :: 3433 : 1573 : 831 : 485 \text{ approx.} \ldots (1).$$

Pauli has calculated the intensity of each of the lines of the Balmer series.

For the line where the transition is $l \rightarrow 2$, he finds the total intensity

$$I_l = \frac{2^6 (l-2)^{2l-3}}{l(l+2)^{2l+3}} (3l^2 - 4)(5l^2 - 4).$$

Taking $l = 3, 4, 5, 6$, the I's for $H_\alpha, H_\beta, H_\gamma, H_\delta$ agree well with (1) above.

Recently Foster[3] has given a series of experimental results on the Stark effect in the arc spectrum of helium, and has calculated the displacements and intensities in fields ranging from 10 to 100 kv./cm. by the use of the Heisenberg perturbation theory of §§ 60 to 63. A good agreement was found.

[1] 'Intensities of Spectral Lines,' by H. A. Kramers, Copenhagen, p. 287, **1919**.

[2] J. Stark, Ann. der Phys. **43**, p. 1004, **1914**.

[3] J. S. Foster, Proc. Roy. Soc. A **114**, p. 47, **1927**, and A **117**, p. 137, **1927**.

CHAPTER XXII

SCHRÖDINGER'S DISPERSION THEORY

111. *Schrödinger's dispersion theory.*

Schrödinger[1] considers the incidence of linearly polarised monochromatic light of frequency ν upon the atom.

If the light is polarised along the z axis so that this is the direction of the electric light-vector $F \cos 2\pi\nu t$, the potential energy of the atom electrons is

$$\begin{aligned} & - F \cos 2\pi\nu t \text{ (total moment along } Oz) \\ & = - F \cos 2\pi\nu t \ (\Sigma ez \text{ for the electrons}) \\ & = A\,(q) \cos 2\pi\nu t, \end{aligned}$$

where A is a function of q, and q stands for all the coordinates of the atom.

Hence $V = V_0\,(q) + A\,(q) \cos 2\pi\nu t$, where V_0 is the potential energy of the undisturbed atom.

The wave equation is

$$\Delta\psi + \frac{8\pi^2}{h^2}(E - V)\,\psi = 0,$$

wherein it is assumed that ψ has a time factor $e^{\frac{2\pi i E t}{h}}$, so that

$$\frac{\partial\psi}{\partial t} = \frac{2\pi i E}{h}\psi.$$

Eliminating E from the wave equation, we have

$$\Delta\psi - \frac{4\pi i}{h} \cdot \frac{\partial\psi}{\partial t} - \frac{8\pi^2 V}{h^2} \cdot \psi = 0.$$

Writing in the above value of V, we obtain

$$\Delta\psi - \frac{4\pi i}{h}\frac{\partial\psi}{\partial t} - \frac{8\pi^2}{h^2}(V_0 + A \cos 2\pi\nu t)\,\psi = 0 \ldots (1).$$

[1] E. Schrödinger, Ann. der Phys. **81**, p. 109, **1926**. Cf. O. Klein, Zs. f. Phys. **41**, p. 407, § 5, **1927**.

For the undisturbed atom

$$\Delta\psi - \frac{4\pi i}{h}\frac{\partial\psi}{\partial t} - \frac{8\pi^2 V_0}{h^2}\psi = 0,$$

the eigen functions and werte are supposed known for it; let them be

$$\psi_1 e^{\frac{2\pi i E_1 t}{h}}, \phi_2 e^{\frac{2\pi i E_2 t}{h}}, \dots \text{ and } E_1, E_2, \dots.$$

Consider the perturbed state of an atom whose undisturbed state is defined by $\psi_n e^{\frac{2\pi i E_n t}{h}}$ and E_n.

Let the ψ_n' (of the perturbation theory of § 106) be written

$$\psi_n' = \psi_n(q)\, e^{\frac{2\pi i E_n t}{h}} + v(q, t),$$

where v is of order A.

Substitute in (1), which is satisfied by ψ_n', and neglect squares of A. Then

$$\Delta v - \frac{4\pi i}{h}\frac{\partial v}{\partial t} - \frac{8\pi^2}{h} V_0 v = \frac{8\pi^2}{h^2} A \cos 2\pi\nu t\, \psi_n e^{\frac{2\pi i E_n t}{h}}$$

$$= \frac{4\pi^2}{h^2} A\psi_n \{e^{\frac{2\pi i t}{h}(E_n + h\nu)} + e^{\frac{2\pi i t}{h}(E_n - h\nu)}\} \dots\dots (2).$$

Write

$$v(q, t) = v_+(q)\, e^{\frac{2\pi i t}{h}(E_n + h\nu)} + v_-(q)\, e^{\frac{2\pi i t}{h}(E_n - h\nu)} \dots (3).$$

Substitution in (2) gives the two equations:

$$\Delta v_\pm + \frac{8\pi^2}{h^2}(E_n \pm h\nu - V_0)\, v_\pm = \frac{4\pi^2}{h^2} A\psi_n \quad \dots (4),$$

where the upper or lower signs are to be taken together.

[The effect of the exciting disturbance A thus depends not only upon A but upon the state (ψ_n) of the atom when A begins to act; this is very different from the ordinary theory of forced vibrations of a string or a plate.]

Equation (4) is

$$\left\{-\frac{h^2}{8\pi^2}\Delta + V_0 - (E_n \pm h\nu)\right\} v_{\pm} = -\tfrac{1}{2}A\psi_n,$$

and in this form corresponds to (4), § 106, the operator $V_0 - \frac{h^2}{8\pi^2}\Delta$ being the equivalent of H and $E_n \pm h\nu$ replacing the E_s.

Using (8), § 106, we have

$$v_{\pm} = \tfrac{1}{2}\Sigma_k{}' \frac{\psi_k \int A\bar{\psi}_k \psi_n \, dq}{E_n \pm h\nu - E_k},$$

where the dash above the Σ excludes from the sum the case $|E_n - E_k| = h\nu$.

If $A\,(kn)$ is the amplitude of the component of the matrix A corresponding to the frequency

$$\nu\,(kn) = (E_k - E_n)/h,$$

then

$$A\,(kn) = \int A\bar{\psi}_k \psi_n \, dq \quad (\S\ 102\ (2));$$

$$\therefore\ v_{\pm} = \tfrac{1}{2}\Sigma_k{}' \frac{A\,(kn)\,\psi_k}{E_n - E_k \pm h\nu} \quad \text{............(5).}$$

Hence

$$\begin{aligned}\psi_n{}' &= \psi_n e^{\frac{2\pi i E_n t}{h}} + v\,(q, t)\\ &= \psi_n e^{\frac{2\pi i E_n t}{h}} + v_+(q)\, e^{\frac{2\pi i t}{h}(E_n + h\nu)} + v_-(q)\, e^{\frac{2\pi i t}{h}(E_n - h\nu)}\\ &= \psi_n e^{\frac{2\pi i E_n t}{h}} + \tfrac{1}{2}\Sigma_k{}' A\,(kn)\,\psi_k \left\{\frac{e^{\frac{2\pi i t}{h}(E_n + h\nu)}}{E_n - E_k + h\nu} + \frac{e^{\frac{2\pi i t}{h}(E_n - h\nu)}}{E_n - E_k - h\nu}\right\} \quad \text{......(6).}\end{aligned}$$

We now follow the procedure of § 103 and calculate the 'density' ρ given by $\rho = \psi_n{}'\bar{\psi}_n{}'$.

The result is

$$\rho = \psi_n \bar{\psi}_n + 2 \cos 2\pi\nu t \Sigma_k' \frac{(E_n - E_k) A (kn) \bar{\psi}_k \psi_n}{(E_n - E_k)^2 - h^2 \nu^2} \quad \ldots(7).$$

Let the dipole moments of the undisturbed atom along y and z, viz. Σey, Σez, have matrix components $b\,(kn)$, $c\,(kn)$, so that

$$\left.\begin{aligned} b\,(kn) &= \int (\Sigma ey)\, \bar{\psi}_k \psi_n \, dq \\ c\,(kn) &= \int (\Sigma ez)\, \bar{\psi}_k \psi_n \, dq \end{aligned}\right\} \quad \ldots\ldots\ldots\ldots\ldots(8).$$

Then, since at the outset of this article, $A = -F\,(\Sigma ez)$, where F is the incident light vector,

$$A\,(kn) = -F.c\,(kn) \quad \ldots\ldots\ldots\ldots\ldots\ldots(9).$$

To find the dipole moment along y for the *disturbed* atom we calculate

$$\int \Sigma ey \rho \, dq$$

$$= \int (\Sigma ey) \left\{ \psi_n^2 + 2 \cos 2\pi\nu t \, \Sigma_k' \frac{(E_n - E_k) A\,(kn) \bar{\psi}_k \psi_n}{(E_n - E_k)^2 - h^2\nu^2} \right\} dq$$

$$= b\,(nn) + 2 \cos 2\pi\nu t \, \Sigma_k' \frac{(E_n - E_k) A\,(kn)\, b\,(kn)}{(E_n - E_k)^2 - h^2\nu^2},$$

using (8),

$$= b\,(nn) - 2F \cos 2\pi\nu t \, \Sigma_k' \frac{(E_n - E_k)\, b\,(kn)\, c\,(kn)}{(E_n - E_k)^2 - h^2\nu^2} \quad \ldots(10).$$

This second term gives the secondary radiation to which the original wave $F \cos 2\pi\nu t$ gives rise; the first term is the constant dipole moment which is eventually connected up with the original free vibrations.

This dispersion formula is identical with that of Kramers and Heisenberg[1] for the secondary radiation, found by correspondence principle methods. It contains the so-called 'negative' terms which correspond to the possi-

[1] H. A. Kramers and W. Heisenberg, Zs. f. Phys. 31, p. 693, equation (29), 1925. See also Q.T.A. §§ 148–152.

bility of changes to a deeper level ($E_k < E_n$) to which Kramers[1] called special attention.

The important connection of the coefficients of the secondary radiation from the atom with its spontaneous emission coefficients $b\,(kn)$, $c\,(kn)$ is brought fully into evidence.

112. *For excited atoms* there may be not just the one eigenfunction ψ_n in the original state, but several, say for example two ψ_n, ψ_m. In the expression for $\psi'\bar{\psi}'$ and for the electric moment of the disturbed atom, not only the terms already found arising directly from the ψ_n and ψ_m occur, but combination terms

$$\psi_n \bar{\psi}_m e^{\frac{2\pi i}{h}(E_n - E_m)t}.$$

These give not only the spontaneous radiation due to the existence of the levels E_n, E_m, but also a perturbation term of the first order, proportional to F, due to the interaction of the ψ_n forced vibrations with the ψ_m free vibration and of the ψ_m forced vibrations with the ψ_n free vibration. The frequency of these new terms in $\psi'\bar{\psi}'$ is seen without calculation to be

$$\left| \nu \pm \frac{E_n - E_m}{h} \right|.$$

These scattered non-coherent waves were found by Kramers and Heisenberg[2], and also predicted by Smekal by the use of light quanta.

[1] H. A. KRAMERS, Nature, **113**, p. 673, and **114**, p. 310, **1924**.
[2] Q.T.A. §§ 153, 154.

CHAPTER XXIII

LIGHT QUANTA; DE BROGLIE WAVES; DIFFRACTION OF LIGHT QUANTA; THE COMPTON EFFECT

113. *Light quanta.*

The history of the corpuscular and wave controversy in optical theory is well known. At the end of the last century physicists had no doubt as to the wave character of light, and the acceptance of the wave theory meant the rejection of the corpuscular theory. The possibility of both being correct and helpful to one another had not been much thought about and still less explored.

At about this time Planck's quantum theory appeared and was soon followed in 1905 by Einstein's theory of 'light quanta' to explain the photo-electric effect[1], for which the wave theory alone had utterly failed to account. Einstein assumed that radiation exists in discrete 'quanta' of energy $h\nu$, where ν is the frequency; thus a quantum is a separate thing, like a corpuscle, and has a frequency associated with it, like a wave. There was a blend of the two older theories involved. This idea has been generalised by the speculations of Louis de Broglie, who pointed out the important clue given by the theory of relativity, namely, that of the identity of mass and energy, so that the conservation of mass is therefore also the conservation of energy. This suggested to him that all forms of energy (including radiation) should have an atomic structure, like matter; and that the atoms of energy are grouped round certain singular points, forming electrons, light quanta, and the like.

[1] Q.T.A. §§ 24, 25.

114. *de Broglie's*[1] *theory of atoms and their associated waves.*

He considers a singular point, such as a material particle or a light quantum. It will have a mass m_0, its 'proper mass' as measured by an observer moving with it. The theory of relativity tells us that the energy associated with it and measured by this observer is m_0c^2, where c is the velocity of light. Consider two Galilean systems of axes, one moving with the particle and the other in which it has a velocity $v \equiv \beta c$ parallel to Oz.

The energy, being m_0c^2 in the former, is $\dfrac{m_0c^2}{\sqrt{1-\beta^2}}$ in the latter.

The difference of these two expressions,

$$m_0c^2\left\{\frac{1}{\sqrt{1-\beta^2}}-1\right\},$$

is the kinetic energy of the point in the second system, and if β is small reduces to the classical value $\frac{1}{2}m_0v^2$.

de Broglie now makes the following fundamental postulate: 'If a material element in the most general sense (electron, proton, light quantum) has energy W, there exists in this system a periodic phenomenon of frequency ν defined by $W = h\nu$.'

Thus the frequency associated with the mass m_0 in the first system will be $\nu_0 = \dfrac{m_0c^2}{h}$ and will be that of a periodic phenomenon spread round the mass point, of which the latter is a singularity, just as an electron is for its electrostatic field. This phenomenon must be analogous to a stationary wave and can be represented by

$$f(x_0, y_0, z_0)\sin 2\pi\nu_0 t_0,$$

where x_0, y_0, z_0, t_0 are the space-time coordinates proper to the first system and $f(x_0, y_0, z_0)$ is the amplitude at each point of the phenomenon.

[1] Louis de Broglie, 'Ondes et mouvements,' Gauthier-Villars, Paris, **1926**.

How would this phenomenon appear to an observer in the second system who sees the particle pass with velocity $v\ (\equiv \beta c)$?

The Lorentz transformation shows that it would appear as

$$f\left(x, y, \frac{z - vt}{\sqrt{1-\beta^2}}\right) \sin \frac{2\pi\nu_0}{\sqrt{1-\beta^2}}\left(t - \frac{\beta z}{c}\right) \quad\text{......(1).}$$

Writing $$\nu = \frac{\nu_0}{\sqrt{1-\beta^2}}, \quad V = \frac{c}{\beta},$$

(1) becomes $$f\left(x, y, \frac{z - vt}{\sqrt{1-\beta^2}}\right) \sin 2\pi\nu\left(t - \frac{z}{V}\right),$$

and represents a wave of frequency ν, whose *amplitude* travels along z with speed v, and whose *phase* travels along z with speed V.

As the energy E of the particle in the x, y, z system is $\frac{m_0 c^2}{\sqrt{1-\beta^2}}$, we have

$$E = \frac{m_0 c^2}{\sqrt{1-\beta^2}} = \frac{h\nu_0}{\sqrt{1-\beta^2}} = h\nu,$$

so that the quantum relation is conserved.

Further, regarding ν, V as functions of β,

$$d\nu = \frac{\nu_0 \beta\, d\beta}{(1-\beta^2)^{\frac{3}{2}}} \text{ and } d\left(\frac{1}{V}\right) = \frac{1}{c}\, d\beta,$$

so that $$\frac{d}{d\nu}\left(\frac{1}{V}\right) = \frac{(1-\beta^2)^{\frac{3}{2}}}{\nu_0 \beta c} \quad\text{..................(2).}$$

$$\therefore \frac{d}{d\nu}\left(\frac{\nu}{V}\right) = \frac{1}{V} + \nu\frac{d}{d\nu}\left(\frac{1}{V}\right)$$

$$= \frac{\beta}{c} + \frac{\nu_0}{\sqrt{1-\beta^2}}\frac{(1-\beta^2)^{\frac{3}{2}}}{\nu_0 \beta c}, \text{ using (2)}$$

$$= \frac{\beta}{c} + \frac{1-\beta^2}{\beta c} = \frac{1}{\beta c} = \frac{1}{v}.$$

Thus $\frac{1}{v} = \frac{d}{d\nu}\left(\frac{\nu}{V}\right)$, which is Rayleigh's relation for dispersive media, connecting the velocity (v) of a group of waves of frequency ν with the velocity (V) of the waves themselves.

Thus the particle is surrounded by a group of waves keeping pace with it with velocity $v = \beta c$, whose individual waves (phases) have a velocity $V = c/\beta$ (and thus $> c$).

115. *The momentum of a light quantum.*

Since $h\nu = E = \frac{m_0 c^2}{\sqrt{1-\beta^2}}$ and the momentum $= mv$, where $m = \frac{m_0}{\sqrt{1-\beta^2}}$, the momentum

$$= \frac{m_0}{\sqrt{1-\beta^2}} v = \frac{h\nu}{c^2} v.$$

For a light quantum $m_0 \to 0$ and $v \to c$, the energy remaining finite and $= h\nu$; the momentum $\to \frac{h\nu}{c}$.

Thus the light quantum has energy $h\nu$ and momentum $\frac{h\nu}{c}$.

116. *The pressure of radiation.*

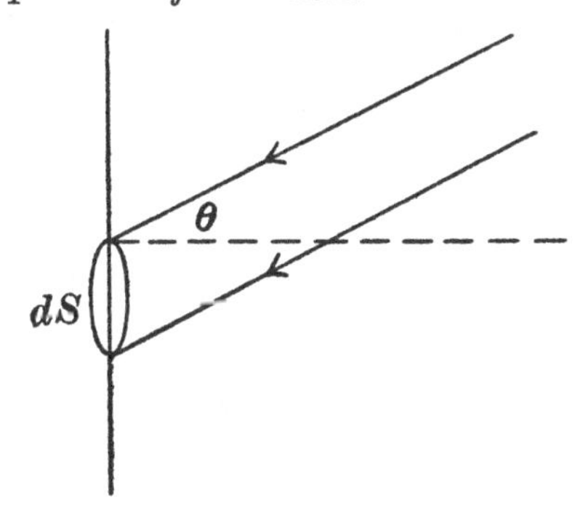

Consider the incidence of monochromatic radiation upon a plane reflector. Let there be n quanta per unit volume, so that $\rho = nh\nu$ is the energy density of the incident beam. Consider an area dS of the reflector. In time dt, the reflector receives $n\,(dS \cos\theta)\,(c\,dt)$ quanta and reflects them.

Before and after reflexion each quantum has momentum $h\nu/c$ and, while the impact does not affect the tangential component, it reverses the normal one. Thus the impulse on the reflector in time δt

$$= 2\,[n\,dS \cos\theta\, c\,dt]\left[\frac{h\nu}{c}\cos\theta\right]$$

or the force on the area $dS = 2n\,dS\,h\nu \cos^2\theta$ or the radiation pressure $= 2nh\nu \cos^2\theta$.

If E is the total energy per unit volume due to the incident and reflected beams, $E = 2\rho = 2nh\nu$, and thus the radiation pressure $p = E \cos^2\theta$.

If we have an infinite number of rays of the same density such that all angles of incidence are equally represented, $p = E\,\overline{\cos^2\theta}$, where the bar denotes the mean value in the range $\theta = 0$ to π, so that $p = E/3$.

117. *Diffraction of light quanta.*

Duane[1] considers a crystal, represented by a space-lattice along whose axes q_1, q_2, q_3 are lattice points at intervals a_1, a_2, a_3. Through the impact of a light quantum, the crystal receives an impulse whose components are p_1, p_2, p_3. But the impulse it can receive is limited to special values given by the quantum conditions $\int p_1\,dq_1 = n_1 h$, etc., where n_1, n_2, n_3 are integers so that

$$n_1 h = p_1 \int dq_1 = p_1 a_1,$$

since the integrals are taken through one period of the lattice.

$$\therefore\; p_1 = \frac{n_1 h}{a_1}, \text{ etc.}$$

[1] W. Duane, Proc. Nat. Acad. Amer. **9**, p. 158, **1923**.

If l, m, n are the direction cosines for the light quantum before impact and l', m', n' those after, then the loss of momentum of the light quantum is $(l-l')h\nu/c$, etc., neglecting the change in ν (§ 115). These must be p_1, p_2, p_3, the momentum given to the crystal.

$$\therefore \quad (l-l')\,h\nu/c = n_1 h/a_1, \text{ etc.}$$

$$\therefore \quad l-l' = n_1 c/a_1 \nu, \text{ etc.}$$

But the Laue[1] formula of the classical diffraction theory is $l-l' = \dfrac{m_1\lambda}{a_1}$, etc., where m_1, m_2, m_3 are the 'order-numbers' of the diffracted rays; since $\lambda = c/\nu$, these are in complete agreement with the Laue m's regarded as quantum numbers.

Jordan[2] has extended the theory to the impact of material particles on a lattice and finds the same results if λ is taken to be the de Broglie wave length associated with the particle.

118. *The Compton effect*[3].

J. J. Thomson in his experiments on the scattering of X-rays found that for a thin layer of matter the intensity of the scattered radiation was the same on the two sides of the plate. Compton found this to be true for moderately hard rays, but for hard or γ rays, the scattered energy was less than the Thomson value and was concentrated on the emergent side of the plate.

On the classical theory each X-ray affects every electron of the plate and the scattering is supposed due to the combined effect of all the electrons; the result is utterly at variance with recent experiment.

On the light quantum theory, used by Compton, any one quantum of X-rays is not scattered by all the electrons

[1] M. Laue, Ann. der Phys. **44**, p. 1197, **1914**.
[2] P. Jordan, Zs. f. Phys. **37**, p. 376, **1926**.
[3] A. H. Compton, Phys. Rev. **21**, p. 483, **1923**.

but by some one electron. He considers the impact of an X-ray quantum of energy $h\nu$ and momentum $h\nu/c$ with an electron at rest.

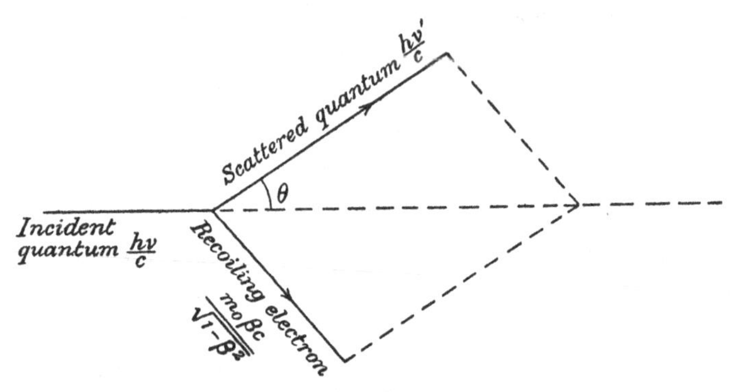

Momentum Diagram

The energy and momentum of the scattered quantum are $h\nu'$, $h\nu'/c$ and those of the electron[1] after impact

$$m_0 c^2 \left[\frac{1}{\sqrt{1-\beta^2}} - 1\right], \qquad \frac{m_0 \beta c}{\sqrt{1-\beta^2}},$$

where βc is the velocity given to the electron.

Using momentum and energy principles we have

$$\left.\begin{aligned} \left(\frac{m_0\beta c}{\sqrt{1-\beta^2}}\right)^2 &= \left(\frac{h\nu}{c}\right)^2 + \left(\frac{h\nu'}{c}\right)^2 - \frac{2h\nu}{c}\cdot\frac{h\nu'}{c}\cdot\cos\theta \\ \text{and} \qquad h\nu &= h\nu' + m_0 c^2\left(\frac{1}{\sqrt{1-\beta^2}} - 1\right) \end{aligned}\right\}.$$

These equations determine β, ν'.

From the second,

$$\frac{1}{1-\beta^2} = \left\{1 + \frac{h(\nu-\nu')}{m_0 c^2}\right\}^2.$$

$$\therefore \quad \frac{\beta^2}{1-\beta^2} = \frac{2h(\nu-\nu')}{m_0 c^2} + \frac{h^2(\nu-\nu')^2}{m_0{}^2 c^4}.$$

[1] Q.T.A. § 36.

Substituting in the first, we have

$$2m_0h(\nu-\nu')+\frac{h^2}{c^2}(\nu-\nu')^2=\frac{h^2}{c^2}(\nu^2+\nu'^2-2\nu\nu'\cos\theta)$$

or
$$2m_0h(\nu-\nu')=\frac{2h^2}{c^2}\nu\nu'(1-\cos\theta)$$

or
$$\frac{1}{\nu'}-\frac{1}{\nu}=\frac{2h}{m_0c^2}\sin^2\frac{\theta}{2}.$$

If λ, λ' are the corresponding wave lengths,

$$\lambda=c/\nu,\quad \lambda'=c/\nu',$$

and therefore

$$\lambda'=\lambda+\frac{2h}{m_0c}\sin^2\frac{\theta}{2}.\qquad\left[\frac{2h}{m_0c}=4{\cdot}8\times10^{-10}\right]\quad\text{...(1).}$$

Hence
$$\frac{\nu}{\nu'}=1+\frac{2h\nu}{m_0c^2}\sin^2\frac{\theta}{2};$$

and writing
$$\alpha=\frac{h\nu}{m_0c^2},$$

we have
$$\frac{\nu}{\nu'}=1+2\alpha\sin^2\frac{\theta}{2},$$

and therefore

$$\frac{1}{\sqrt{1-\beta^2}}=1+\frac{h(\nu-\nu')}{m_0c^2}=1-\alpha\left(1-\frac{\nu'}{\nu}\right).$$

$$\therefore\ \beta=\frac{2\alpha\sin\frac{\theta}{2}\sqrt{1+(2\alpha+\alpha^2)\sin^2\frac{\theta}{2}}}{1+(2\alpha+\alpha^2)\sin^2\frac{\theta}{2}}\quad\text{...(2).}$$

Equation (1) shows that $\lambda'>\lambda$, the increase ranging from a few per cent. for ordinary X-rays to more than 200 per cent. in the case of γ rays scattered backwards ($\theta=\pi$). At the same time the velocity of recoil (βc) varies from 0 when the ray is scattered directly forward to about 80 per cent. of c when a γ ray is scattered at a large angle ($\theta\sim\pi$).

From (1) and (2) Compton found that for incident radiation of intensity I_0, the intensity $I(r,\theta)$ of the radiation scattered in a direction θ by N electrons had at a distance r the value given by

$$I(r,\theta) = I_0 \frac{Ne^4}{2r^2m^2c^4} \cdot \frac{1+\cos^2\theta + 2\alpha(1+\alpha)(1-\cos\theta)^2}{\{1+\alpha(1-\cos\theta)\}^5} \quad \text{............(3).}$$

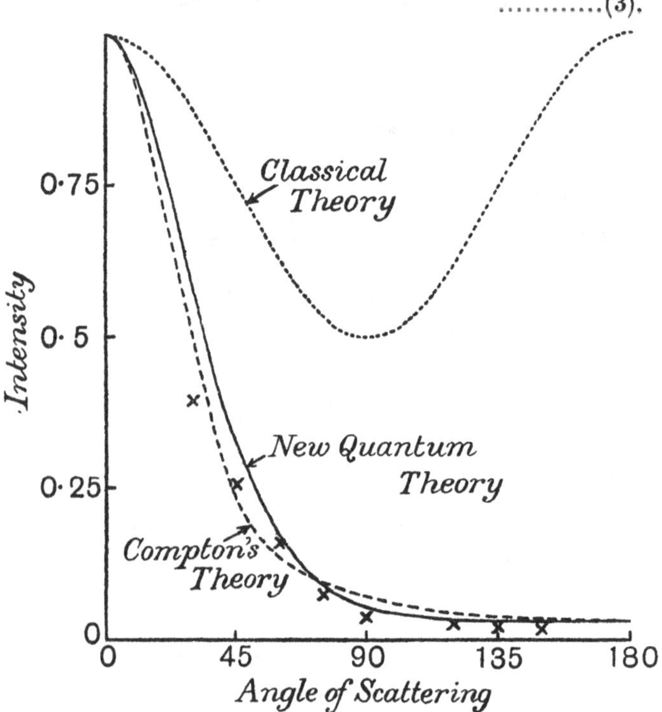

Dirac[1] has lately calculated the Compton effect by the use of his new relativity quantum mechanics (§ 104). He finds that for plane polarised incident radiation of intensity I_0, the scattered radiation at a distance r in a direction whose polar angles are θ, ϕ is

$$I(r,\theta,\phi) = I_0 \frac{Ne^4}{r^2m^2c^4} \cdot \frac{\sin^2\phi}{\{1+\alpha(1-\cos\theta)\}^3},$$

[1] P. A. M. Dirac, Proc. Roy. Soc. A. 111, p. 405, 1926.

where θ is the angle of scattering and ϕ is the angle between the direction of the scattered radiation and the direction of the electric vector of the incident radiation.

For unpolarised incident radiation, Dirac deduces

$$I(r, \theta) = I_0 \frac{Ne^4}{2r^2m^2c^4} \cdot \frac{1 + \cos^2\theta}{\{1 + \alpha(1 - \cos\theta)\}^3} \quad \ldots(4).$$

The full curve of the figure shows the variation with θ of the intensity of the scattered radiation according to Dirac's formula (4), for unpolarised incident radiation of wave length ·022 Å (and therefore $\alpha = 1{\cdot}2$). The lower broken curve gives the results of Compton's light quantum theory and represents his formula (3); the upper dotted curve is the result of the classical theory, which expresses $I(r, \theta)$ as

$$I_0 \frac{Ne^4}{2r^2m^2c^4}(1 + \cos^2\theta).$$

The crosses indicate Compton's experimental values.

See also W. Bothe and H. Geiger, Zs. f. Phys. **32**, p. 639, **1925**.

CHAPTER XXIV

THE THEORY OF THE ANOMALOUS ZEEMAN EFFECT ON THE NEW MECHANICS

119. *The precession of a spinning electron moving through an electric field.*

Owing to the motion of the electron through the electric field $\mathbf{E}$ with velocity $\mathbf{v}$ it is subject to a magnetic field $\mathbf{H} = \frac{1}{c}(\mathbf{E}_\wedge\mathbf{v})$, where $\mathbf{E}_\wedge\mathbf{v}$ means the *vector* product of $\mathbf{E}$ and $\mathbf{v}$.

If $\mathbf{M}$ is the magnetic moment and $\mathbf{S}$ the angular momentum of the spin of the electron, $\mathbf{M} = \frac{e}{m_0 c}\mathbf{S}$. (§ 22.)

The effect of the magnetic field $\mathbf{H}$ is to cause the spin axis to precess about $\mathbf{H}$ with angular velocity $\boldsymbol{\omega}$. The energy ΔE due to this precession is $(\boldsymbol{\omega}.\mathbf{S})$, where $\boldsymbol{\omega}.\mathbf{S}$ means the *scalar* product of $\boldsymbol{\omega}$ and $\mathbf{S}$ (§ 7); and it is also $\mathbf{H}.\mathbf{M}$ by the usual formula for the energy of a magnet in a magnetic field. Thus $\omega.\mathbf{S} = \mathbf{H}.\mathbf{M} = \frac{e}{m_0 c}\mathbf{H}.\mathbf{S}$.

$$\therefore \quad \omega = \frac{e\mathbf{H}}{m_0 c},$$

just twice the Larmor value.

Substituting for $\mathbf{H}$, we have

$$\boldsymbol{\omega} = \frac{e}{m_0 c^2}(\mathbf{E}_\wedge\mathbf{v}) \quad \ldots\ldots\ldots\ldots\ldots\ldots(1).$$

When this value was applied by Heisenberg and Pauli to calculate the width of the spin doublets, the value found was twice that of experiment.

120. *Thomas's[1] relativity correction of this result.*

Thomas resolved this difficulty by pointing out that the precession calculated in § 119 is the precession observed

[1] L. H. Thomas, Nature, **117**, p. 514, **1926**; Phil. Mag. **3**, p. 1, **1927**. See also J. Frenkel, Zs. f. Phys. **37**, p. 243, **1926**.

in a coordinate system in which the electron is at rest; whereas what is needed is the precession observed in a system in which the nucleus is at rest and the electron moves with velocity **v**.

Let 1 denote a coordinate system in which the nucleus is at rest at time t, 2 a system in which the electron is at rest at time t, and 3 a system in which the electron is at rest at time $t + dt$.

Relative to 1, the velocity of the electron in 2 is **v** and the velocity of the electron in 3 is $\mathbf{v} + \mathbf{f}dt$, where **f** is the acceleration of the electron.

The system 2 is derived from 1 by a Lorentz transformation with velocity **v**, and 3 from 1 by a Lorentz transformation with velocity $\mathbf{v} + \mathbf{f}dt$.

The precession seen by an observer in 1 (which is what is sought) is that precession which would turn the direction of the spin axis at time t in 2 to its direction at time $t + dt$ in 3, both directions being regarded as directions in 1. To a first approximation 3 is found from 2 by a Lorentz transformation with velocity $\mathbf{f}dt$ together with a rotation $\frac{1}{2c^2}(\mathbf{v}_\wedge \mathbf{f})\, dt$.

Thus to a first approximation, the observed rate of precession is

$$\frac{e}{m_0 c^2}(\mathbf{E}_\wedge \mathbf{v}) - \frac{1}{2c^2}(\mathbf{v}_\wedge \mathbf{f}),$$

and writing $\mathbf{f} = -\frac{\mathbf{E}e}{m_0}$ to this order, the rate becomes

$$\frac{e}{m_0 c^2}(\mathbf{E}_\wedge \mathbf{v}) + \frac{1}{2c^2}\left(\mathbf{v}_\wedge \frac{\mathbf{E}e}{m_0}\right)$$

$$= \frac{e}{m_0 c^2}(\mathbf{E}_\wedge \mathbf{v}) - \frac{e}{2m_0 c^2}(\mathbf{E}_\wedge \mathbf{v})$$

$$= \frac{1}{2}\frac{e}{m_0 c^2}(\mathbf{E}_\wedge \mathbf{v}) \quad \ldots\ldots\ldots\ldots\ldots\ldots\ldots\ldots (2),$$

which is just *half* of (1). This is Thomas's result.

For a Coulomb field where the nuclear charge is Ze,

$$\mathbf{E} = \frac{Ze\mathbf{r}}{r^3}$$

and the precession (2) becomes

$$\frac{Ze^2}{2m_0c^2r^3}(\mathbf{r}_\wedge \mathbf{v}).$$

But the angular momentum $\mathbf{L}$ in the orbit is $m_0(\mathbf{r}_\wedge \mathbf{v})$; so that the precession $\boldsymbol{\omega}$ is

$$\frac{Ze^2}{2m_0^2c^2r^3}\mathbf{L}.$$

Hence the energy term due to the precession, being $(\boldsymbol{\omega}.\mathbf{S})$ (§ 7), is

$$\frac{Ze^2}{2m_0^2c^2r^3}(\mathbf{L}.\mathbf{S}) \quad \text{..................(3)}.$$

Finally, the magnetic field $\mathbf{H} \equiv (\mathbf{E}_\wedge \mathbf{v})$ is perpendicular to each of the vectors $\mathbf{E}$, $\mathbf{v}$ and therefore in the Coulombian case is normal to the plane of the orbit. Thus the precession of the spinning electron is about the normal to the plane of the orbit.

121. *Relativity correction for the Keplerian orbit.*

The Hamiltonian[1] for an electron moving about a nuclear charge Ze is

$$H = c\sqrt{m_0^2c^2 + p_x^2 + p_y^2 + p_z^2} - m_0c^2 - \frac{Ze^2}{r}$$

$$= m_0c^2\left\{\frac{\frac{1}{2}(p_x^2 + p_y^2 + p_z^2)}{m_0^2c^2} - \frac{\frac{1}{8}(p_x^2 + p_y^2 + p_z^2)^2}{m_0^4c^4}\cdots\right\} - \frac{Ze^2}{r}$$

$$= \underbrace{\frac{p_x^2 + p_y^2 + p_z^2}{2m_0} - \frac{Ze^2}{r}}_{H_0} - \underbrace{\frac{(p_x^2 + p_y^2 + p_z^2)^2}{8m_0^3c^2}}_{H_1}, \text{ approx.}$$

We regard the relativity term H_1 as a perturbation of the usual orbit given by H_0.

[1] Q.T.A. p. 80.

The additional term to the energy due to the perturbation is the mean value $\bar{H}_1$ of H_1 taken over the undisturbed path[1] (§ 61).

If E_0 is the energy in the undisturbed orbit,

$$\frac{p_x^2 + p_y^2 + p_z^2}{2m_0} - \frac{Ze^2}{r} = E_0,$$

so that
$$H_1 = -\frac{\left\{2m_0\left(E_0 + \frac{Ze^2}{r}\right)\right\}^2}{8m_0^3c^2}.$$

$$\therefore\ \bar{H}_1 = -\frac{1}{2m_0c^2}\left\{E_0^2 + 2Ze^2E_0\left(\overline{\frac{1}{r}}\right) + Z^2e^4\left(\overline{\frac{1}{r^2}}\right)\right\},$$

which is the additional relativity energy term.

122. *Heisenberg and Jordan's calculation*[2] *of the anomalous Zeeman effect by the new mechanics.*

The problem considered is the motion of a spinning electron about a nucleus under the action of a magnetic field.

The electron charge is $-e$, the effective nuclear charge Ze, the angular momentum of the electron due to its spin is **S** and the angular momentum in the orbit is **L**. The behaviour of the electron, neglecting relativity, the magnetic field, and the spin of the electron, has been given on the new mechanics by the Dirac-Pauli theory of the hydrogen atom (§ 81).

The present problem is to find the perturbing effect of these three causes on the Dirac-Pauli result.

The perturbation energy in the Hamiltonian is the sum of three terms.

Term 1 is that due to the action of the magnetic field on the orbit and the electron-magnet. The magnetic

[1] Q.T.A. § 133.
[2] W. Heisenberg and P. Jordan, Zs. f. Phys. **37**, p. 263, **1926**.

moment associated with the former is $\frac{e}{2m_0c}\mathbf{L}$ and the latter $\frac{e}{m_0c}\mathbf{S}$, so that term 1 is

$$\mathbf{H}.\left(\frac{e}{2m_0c}\mathbf{L}\right)+\mathbf{H}.\left(\frac{e}{m_0c}\mathbf{S}\right)=\frac{e}{2m_0c}\mathbf{H}.(\mathbf{L}+2\mathbf{S}).$$

Term 2 is that due to the action of the electric field upon the moving electron-magnet; this has been found (§ 120) to be $\frac{Ze^2}{2m_0{}^2c^2r^3}(\mathbf{L}.\mathbf{S})$.

Term 3 is due to relativity mass variation and is given by § 121.

If the *mean values* of these terms are taken over the undisturbed path and are denoted by $\bar{H}_1$, $\bar{H}_2$, $\bar{H}_3$, then

$$\bar{H}_1=\frac{e}{2m_0c}\overline{\mathbf{H}.(\mathbf{L}+2\mathbf{S})},$$

$$\bar{H}_2=\frac{Ze^2}{2m_0{}^2c^2}\overline{\frac{1}{r^3}(\mathbf{L}.\mathbf{S})},$$

$$\bar{H}_3=-\frac{1}{2m_0c^2}\left\{E_0{}^2+2Ze^2E_0\left(\overline{\frac{1}{r}}\right)+Z^2e^4\left(\overline{\frac{1}{r^2}}\right)\right\},$$

where the bars denote mean values.

For an alkali atom where there are inner electrons the degeneracy of the hydrogen atom is removed and $\mathbf{L}$, $\mathbf{S}$, M_z may be taken as quantised with the respective sets of values

$$\frac{h}{2\pi}\sqrt{l(l+1)},\quad \frac{h}{2\pi}\sqrt{s(s+1)},\quad \frac{h}{2\pi}m,$$

so that they are diagonal matrices. (M_z is defined in § 72.)

The undisturbed system is then only degenerate in respect of one coordinate. This is S_z, the component of $\mathbf{S}$ along the field, $\equiv \frac{h}{2\pi}.m_s$.

Owing to this degeneracy $\bar{H}_1+\bar{H}_2+\bar{H}_3$ will not be a diagonal matrix and the perturbation theory for degenerate systems (§ 62) must be used, where by a transformation $S(\bar{H}_1 + \bar{H}_2 + \bar{H}_3)S^{-1} = E$, $\bar{H}_1 + \bar{H}_2 + \bar{H}_3$ becomes a diagonal matrix E and so discloses the energy levels. [$\bar{H}_1 + \bar{H}_2 + \bar{H}_3$ is the $\overline{\lambda \bar{H}_1}$ of the theory of § 62.]

123. *The calculations.*

The part $\bar{H}_3$ of the mean perturbation energy is not affected by the transformation, since it contains no degenerate coordinates; it can be added on at the end.

For the momenta **L**, **S** the general rules of §§ 71, 78 apply. Thus $[L_x L_y] = L_z$, etc.; $[S_x S_y] = S_z$, etc.; and each S component commutes with each L component.

Writing $L_z = \frac{h}{2\pi} m_l$ and recalling $S_z = \frac{h}{2\pi} m_s$ (§ 122) we have $m_l + m_s = m$, since **S** and **L** compound into **J** the angular momentum of the whole atom. (**J** is the M of § 71, whose z component $M_z = \frac{h}{2\pi} m$.)

From § 77,

$$\left.\begin{aligned}
(L_x + iL_y)(l, m_l, l, m_l - 1) &= \frac{h}{2\pi}\sqrt{l(l+1) - m_l(m_l - 1)}\\
(L_x - iL_y)(l, m_l - 1, l, m_l) &= \frac{h}{2\pi}\sqrt{l(l+1) - m_l(m_l - 1)}\\
(L_x - iL_y)(l, m_l, l, m_l - 1) &= 0\\
(L_x + iL_y)(l, m_l - 1, l, m_l) &= 0\\
L_x(l, m_l, l, m_l) &= 0\\
L_y(l, m_l, l, m_l) &= 0\\
L_z(l, m_l, l, m_l) &= \frac{h}{2\pi} m_l\\
L_z(l, m_l, l, m_l - 1) &= 0\\
L_z(l, m_l - 1, l, m_l) &= 0
\end{aligned}\right\} \quad (1).$$

The same relations hold with S written for L, s for l, and m_s for m_l ....(2).

Now

$$\overline{H}_1 + \overline{H}_2 = \frac{e}{2m_0c}\,\mathbf{H}.(\mathbf{L} + 2\mathbf{S}) + \frac{Ze^2}{2m_0{}^2c^2}\overline{\left(\frac{1}{r^3}\right)}\,\mathbf{L}.\mathbf{S}.$$

Writing

$$\mu = \frac{e}{2m_0c}\,|\mathbf{H}|\,\frac{h}{2\pi} \quad \text{and} \quad \lambda = \frac{Ze^2}{2m_0{}^2c^2}\overline{\left(\frac{1}{r^3}\right)}\left(\frac{h}{2\pi}\right)^2 \quad \ldots(3),$$

and observing that $\mathbf{H}.\mathbf{L} = |\mathbf{H}|\,L_z$, we have

$$\overline{H}_1 + \overline{H}_2 = \frac{2\pi}{h}\,\mu\,(L_z + 2S_z) + \left(\frac{2\pi}{h}\right)^2 \lambda\,(L_xS_x + L_yS_y + L_zS_z)$$

$$= \frac{2\pi}{h}\,\mu\,(L_z + 2S_z) + \left(\frac{2\pi}{h}\right)^2 \lambda\,\{L_zS_z + \tfrac{1}{2}(L_x + iL_y)(S_x - iS_y) + \tfrac{1}{2}(L_x - iL_y)(S_x + iS_y)\}.$$

The indices m, l, s are constant during the changes of m_s, and corresponding to changes $\Delta m_s = 1$, 0 or -1, there are changes $\Delta m_l = -1$, 0 or 1, since

$$m_l + m_s = m.$$

Hence using relations (1) and (2) we have

$$(\overline{H}_1 + \overline{H}_2)(m_sm_s) = \mu\,(m_l + 2m_s) + \lambda m_l m_s$$
$$= \mu\,(m + m_s) + \lambda m_s\,(m - m_s) \equiv \phi\,(m_s), \text{ suppose} \ldots(4)$$

and

$$(\overline{H}_1 + \overline{H}_2)(m_s, m_s - 1)$$
$$= \left(\frac{2\pi}{h}\right)^2 \frac{\lambda}{2}\,(L_x - iL_y)(m_l, m_l + 1)\,(S_x + iS_y)(m_s, m_s - 1),$$

since $m_s \to m_s - 1$ means also $m_l \to m_l + 1$, and the other terms of the expression vanish for these changes.

$$\therefore\ (\overline{H}_1 + \overline{H}_2)(m_s, m_s - 1)$$
$$= \frac{\lambda}{2}\sqrt{\{l(l + 1) - (m_l + 1)\,m_l\}\,\{s(s + 1) - m_s(m_s - 1)\}}$$
$$= \frac{\lambda}{2}\sqrt{\{l(l+1) - (m - m_s + 1)(m - m_s)\}\,\{s(s+1) - m_s(m_s - 1)\}}$$
$$\equiv \psi\,(m_s), \text{ suppose} \quad \ldots\ldots\ldots\ldots\ldots\ldots\ldots\ldots\ldots\ldots(5).$$

In a similar manner $(\bar{H}_1 + \bar{H}_2)(m_s - 1, m_s)$ is found to be equal to this same $\psi(m_s)$.

The number of possible values of m_s corresponding to a given set of values of l, s, m is fixed by the conditions

$$\left.\begin{array}{r} -s \leqslant m_s \leqslant s \\ -l \leqslant m_l \leqslant l \end{array}\right\},$$

or

$$\left.\begin{array}{r} -s \leqslant m_s \leqslant s \\ m - l \leqslant m_s \leqslant m + l \end{array}\right\} \quad \ldots\ldots\ldots\ldots\ldots (6),$$

using

$$m_l + m_s = m.$$

The different values of m_s correspond to the $n_1, n_2, n_3, \ldots$ of the perturbation theory of § 62 for a degenerate system.

Let m_1 be the minimum and m_2 the maximum value of m_s satisfying the conditions (6).

Then the determinant equation (2) of § 62 becomes here

$$0 = \left| \begin{array}{lll} E-(\bar{H}_1+\bar{H}_2)(m_1, m_1), & -(\bar{H}_1+\bar{H}_2)(m_1+1, m_1), & -(\bar{H}_1+\bar{H}_2)(m_1+2, m_1) \ldots\ldots \\ -(\bar{H}_1+\bar{H}_2)(m_1, m_1+1), & E-(\bar{H}_1+\bar{H}_2)(m_1+1, m_1+1), & -(\bar{H}_1+\bar{H}_2)(m_1+2, m_1+1) \ldots\ldots \\ \ldots\ldots\ldots & \ldots\ldots\ldots & \ldots\ldots\ldots \end{array} \right|$$

or

$$0 = \left| \begin{array}{lllll} E-\phi(m_1), & -\psi(m_1+1), & 0, & 0, & \ldots\ldots\ldots \\ -\psi(m_1+1), & E-\phi(m_1+1), & -\psi(m_1+2), & 0, & \ldots\ldots\ldots \\ 0, & -\psi(m_1+1), & E-\phi(m_1+2), & -\psi(m_1+3), & \ldots 0 \\ \ldots\ldots & \ldots\ldots & \ldots\ldots & \ldots\ldots & \ldots\ldots \end{array} \right| \ldots (7),$$

where the ϕ's and ψ's are given by (4) and (5).

This is an equation of degree $m_2 - m_1 + 1$ for E with coefficients *rational* in l, m, s.

The sum of the roots is

$$\phi(m_1) + \phi(m_1 + 1) + \ldots + \phi(m_2).$$

$$\therefore \sum_{n=m_1}^{n=m_2} E_n = \phi(m_1) + \ldots + \phi(m_2)$$

$$= \Sigma_{m_1}^{m_2} \{\mu(m + m_s) + \lambda m_s (m - m_s)\}.$$

Thus the sum of the energy terms arising from (7) is linear in λ, μ and this expresses the 'summation principle' of the Zeeman effect. This principle is that on passage from a weak to a strong magnetic field the g-sums and the γ-sums remain constant.

The former (for the g-sums) was discovered by Heisenberg[1] and extended by Pauli[2]; the latter (for the γ-sums) is due to Landé[3].

[1] W. HEISENBERG, Zs. f. Phys. **8**, p. 273, **1922**.
[2] W. PAULI, Zs. f. Phys. **16**, p. 155, **1923**.
[3] A. LANDÉ, Zs. f. Phys. **19**, p. 112, **1923**, and § 14 of 'Zeemaneffekt und Multiplettstruktur der Spektrallinien,' by E. BACK and A. LANDÉ, Berlin, **1925**.

CHAPTER XXV

THE CALCULATION OF THE ZEEMAN INTENSITIES FOR THE D-DOUBLET OF THE ALKALIS; DEDUCTION OF THE LANDÉ g AND γ FORMULAE; THE FINE STRUCTURE DUE TO RELATIVITY AND SPIN; THE RELATION OF THE SPINNING ELECTRON TO WAVE MECHANICS

124. *Detailed calculation for an alkali doublet.*

Here $s = \frac{1}{2}$, so that by (6) § 123, m_s must be $\pm \frac{1}{2}$.

If $m_s = \frac{1}{2}$, then $m - l \leqslant \frac{1}{2} \leqslant m + l$, or $m \leqslant l + \frac{1}{2}$ only.

If $m_s = -\frac{1}{2}$, then $m - l \leqslant -\frac{1}{2} \leqslant m + l$, or $m \leqslant l - \frac{1}{2}$ only.

The upper limit for m is the j of § 13, so that corresponding to $m_s = \frac{1}{2}, j = l + \frac{1}{2}$, and to $m_s = -\frac{1}{2}, j = l - \frac{1}{2}$. This agrees with § 23.

The determinant equation (7) reduces to

$$\begin{vmatrix} E - \phi\left(-\frac{1}{2}\right), & -\psi\left(\frac{1}{2}\right) \\ -\psi\left(\frac{1}{2}\right), & E - \phi\left(\frac{1}{2}\right) \end{vmatrix} = 0, \qquad \text{since here } m_1 = -\frac{1}{2}$$

or

$$\begin{vmatrix} E-\mu(m-\frac{1}{2})+\frac{\lambda}{2}(m+\frac{1}{2}), & -\frac{\lambda}{2}\sqrt{\{l(l+1)-(m+\frac{1}{2})(m-\frac{1}{2})\}\{\frac{3}{4}+\frac{1}{4}\}} \\ -\frac{\lambda}{2}\sqrt{\{l(l+1)-(m+\frac{1}{2})(m-\frac{1}{2})\}\{\frac{3}{4}+\frac{1}{4}\}}, & E-\mu(m+\frac{1}{2})-\frac{\lambda}{2}(m-\frac{1}{2}) \end{vmatrix} = 0$$

$$\text{or } \left(E - \mu m + \frac{\lambda}{4}\right)^2 - \tfrac{1}{4}(\mu + \lambda m)^2 = \frac{\lambda^2}{4}\{l(l+1) - m^2 + \tfrac{1}{4}\} = 0$$

$$\text{or } E^2 - \left(2\mu m - \frac{\lambda}{2}\right)E + \mu^2(m^2 - \tfrac{1}{4}) - \mu\lambda m - \frac{\lambda^2}{4}l(l+1) = 0$$

$$\text{or} \qquad E = \mu m - \frac{\lambda}{4} \pm \tfrac{1}{2}\sqrt{\mu^2 + 2\mu\lambda m + \lambda^2(l+\tfrac{1}{2})^2} \quad \ldots(1).$$

This is in exact agreement with the Sommerfeld[1] formula of § 39 obtained by adapting the Voigt coupling formula to the quantum theory.

The μ above is ho and the λ is $\frac{h\omega}{l+\frac{1}{2}}$, where $\omega = \Delta\nu_0$, the original width of the doublet (§ 39).

125. *The intensities of the Zeeman components of the doublet.*

We now use H to denote $\bar{H}_1 + \bar{H}_2$ and carry out the transformation $SHS^{-1} = E$, which led to the energy equation (7) of § 123.

Then $$SH = ES.$$

$$\therefore \quad \Sigma_k S(nk)\,H(km) = \Sigma_k E(nk)\,S(km)$$
$$= E(nn)\,S(nm),$$

since E is a diagonal matrix..................................(1).

[n, k, m denote values of m_s.]

From § 123,

only $\quad H(m_s, m_s),\ H(m_s, m_s - 1),\ H(m_s - 1, m_s)$

have non-zero values and since $m_s = \pm\frac{1}{2}$ for the doublet, only

$$H(\tfrac{1}{2}\,\tfrac{1}{2}),\ H(\tfrac{1}{2} - \tfrac{1}{2}),\ H(-\tfrac{1}{2}\,\tfrac{1}{2}),\ H(-\tfrac{1}{2} - \tfrac{1}{2})$$

have non-zero values.

Hence from (1), writing $n = \frac{1}{2}$, $m = \frac{1}{2}$ and so $k = \pm\frac{1}{2}$,

[1] A. SOMMERFELD, Zs. f. Phys. 8, p. 257, 1922.

$$\left.\begin{array}{l} S(\tfrac{1}{2}\,\tfrac{1}{2})\,H(\tfrac{1}{2}\,\tfrac{1}{2}) + S(\tfrac{1}{2}\,-\tfrac{1}{2})\,H(-\tfrac{1}{2}\,\tfrac{1}{2}) = E_{\frac{1}{2}}\,S(\tfrac{1}{2}\,\tfrac{1}{2}); \\ \text{writing } n = \dfrac{1}{2},\ m = -\dfrac{1}{2} \text{ and so } k = \pm\dfrac{1}{2}, \\ S(\tfrac{1}{2}\,\tfrac{1}{2})\,H(\tfrac{1}{2}\,-\tfrac{1}{2}) + S(\tfrac{1}{2}\,-\tfrac{1}{2})\,H(-\tfrac{1}{2}\,-\tfrac{1}{2}) = E_{\frac{1}{2}}\,S(\tfrac{1}{2}\,-\tfrac{1}{2}) \end{array}\right\}$$

where $E_{\frac{1}{2}} = E(\tfrac{1}{2}\,\tfrac{1}{2})$.

From the second of these, using (4) and (5) § 123, we have

$$S(\tfrac{1}{2}\,\tfrac{1}{2})\left\{\frac{\lambda}{2}\sqrt{l(l+1)-(m+\tfrac{1}{2})(m-\tfrac{1}{2})}\right\}$$
$$= S(\tfrac{1}{2}\,-\tfrac{1}{2})\left\{E_{\frac{1}{2}} - \mu(m-\tfrac{1}{2}) + \frac{\lambda}{2}(m+\tfrac{1}{2})\right\}.$$

$$\left.\begin{array}{l} \therefore\ S(\tfrac{1}{2}\,\tfrac{1}{2}) = C\left\{E_{\frac{1}{2}} - \mu(m-\tfrac{1}{2}) + \dfrac{\lambda}{2}(m+\tfrac{1}{2})\right\} \\ S(\tfrac{1}{2}\,-\tfrac{1}{2}) = C\left\{\dfrac{\lambda}{2}\sqrt{l(l+1)-m^2+\tfrac{1}{4}}\right\} \end{array}\right\} \quad (2),$$

where C is a constant. The value of C is determined by the normalising condition $S\tilde{S}^* = 1$, which here gives

$$S(\tfrac{1}{2}\,\tfrac{1}{2})\,\tilde{S}^*(\tfrac{1}{2}\,\tfrac{1}{2}) + S(\tfrac{1}{2}\,-\tfrac{1}{2})\,\tilde{S}^*(-\tfrac{1}{2}\,\tfrac{1}{2}) = 1$$

or

$$S(\tfrac{1}{2}\,\tfrac{1}{2})\,S^*(\tfrac{1}{2}\,\tfrac{1}{2}) + S(\tfrac{1}{2}\,-\tfrac{1}{2})\,S^*(\tfrac{1}{2}\,-\tfrac{1}{2}) = 1$$

or

$$|S(\tfrac{1}{2}\,\tfrac{1}{2})|^2 + |S(\tfrac{1}{2}\,-\tfrac{1}{2})|^2 = 1.$$

Therefore using (2), we have

$$|C|^2\left[\left\{E_{\frac{1}{2}} - \mu(m-\tfrac{1}{2}) + \frac{\lambda}{2}(m+\tfrac{1}{2})\right\}^2 + \frac{\lambda^2}{4}\{l(l+1)-m^2+\tfrac{1}{4}\}\right] = 1.$$

Substituting for $E_{\frac{1}{2}}$ from (1) § 124, this becomes

$$\frac{4}{|C|^2} = \{\mu + \lambda m + \sqrt{\mu^2 + 2\mu\lambda m + \lambda^2(l+\tfrac{1}{2})^2}\}^2 + \lambda^2\{l(l+1) - m^2 + \tfrac{1}{4}\}.$$

$$\therefore\ \frac{2}{|C|^2} = \mu^2 + 2\mu\lambda m + \lambda^2(l+\tfrac{1}{2})^2 + (\mu+\lambda m)\sqrt{\mu^2 + 2\mu\lambda m + \lambda^2(l+\tfrac{1}{2})^2}$$
$$= \Delta^2 + \Delta(\mu + \lambda m),$$

where $$\Delta \equiv \sqrt{\mu^2 + 2\mu\lambda m + \lambda^2 (l + \tfrac{1}{2})^2}.$$

$$\therefore \ | C |^2 = \frac{2}{(\mu + \lambda m + \Delta)\,\Delta}.$$

Thus we have

$$\left.\begin{aligned} S(\tfrac{1}{2}\,\tfrac{1}{2}) &= \tfrac{1}{2}C(\mu + \lambda m + \Delta) \\ S(\tfrac{1}{2} - \tfrac{1}{2}) &= \tfrac{1}{2}C\lambda\sqrt{(l + \tfrac{1}{2})^2 - m^2} \end{aligned}\right\} \quad \ldots\ldots\ldots(3),$$

where $$| C |^2 = \frac{2}{(\mu + \lambda m + \Delta)\,\Delta}$$

and $$\Delta = \sqrt{\mu^2 + 2\mu\lambda m + \lambda^2 (l + \tfrac{1}{2})^2}.$$

So $$\left.\begin{aligned} S(-\tfrac{1}{2}\,\tfrac{1}{2}) &= \tfrac{1}{2}C'(\mu + \lambda m - \Delta) \\ S(-\tfrac{1}{2} - \tfrac{1}{2}) &= \tfrac{1}{2}C'\lambda\sqrt{(l + \tfrac{1}{2})^2 - m^2} \end{aligned}\right\} \quad \ldots\ldots(4),$$

where $$| C' |^2 = \frac{2}{(-\mu - \lambda m + \Delta)\,\Delta}.$$

The transformation matrix S is thus found.

To find the intensities we have $SqS^{-1} = q'$, where q is an undisturbed function of the coordinates x, y or z, and q' the perturbed value; also q is a diagonal matrix for m_s. It suffices to consider the case where $l \to l - 1$ (cf. § 2); this will be understood, and the l not written in.

From § 80 (replacing j by l and m by m_l)

$$\left.\begin{aligned} (x + iy)(m_l, m_l - 1) &= -B\sqrt{(l + m_l)(l + m_l - 1)} \\ (x - iy)(m_l - 1, m_l) &= B\sqrt{(l - m_l)(l - m_l - 1)} \\ z(m_l m_l) &= B\sqrt{l^2 - m_l^2} \end{aligned}\right\} \ldots(5).$$

The general formulae are rather complicated and we confine ourselves to lines of the sodium D type, where l changes from $1 \to 0$.

Calculation of the Zeeman intensities for the sodium D type of doublet.

Since $-l \leqslant m_l \leqslant l$, m_l is $1, 0, -1$ at the beginning of the transition (where $l = 1$) and is 0 at the end (where $l = 0$).

Since $-s \leqslant m_s \leqslant s$, and $s = \frac{1}{2}$, then $m_s = \pm\frac{1}{2}$.

Now $m = m_s + m_l$, so that the possibilities are at the start

$$m_l = 1, \quad m_s = \frac{1}{2} \text{ or } -\frac{1}{2}, \quad m = \frac{3}{2} \text{ or } \frac{1}{2},$$

$$m_l = 0, \quad m_s = \frac{1}{2} \text{ or } -\frac{1}{2}, \quad m = \frac{1}{2} \text{ or } -\frac{1}{2},$$

$$m_l = -1, \quad m_s = \frac{1}{2} \text{ or } -\frac{1}{2}, \quad m = -\frac{1}{2} \text{ or } -\frac{3}{2},$$

and at the end

$$m_l = 0, \quad m_s = \frac{1}{2} \text{ or } -\frac{1}{2}, \quad m = \frac{1}{2} \text{ or } -\frac{1}{2}.$$

Polarisation	m_l	*Initial* m m_s → *Final* m m_s	
⊥	$1 \to 0$	$\frac{3}{2}\ \frac{1}{2} \to \frac{1}{2}\ \frac{1}{2}$ $\frac{1}{2}\ -\frac{1}{2} \to \frac{1}{2}\ \frac{1}{2}$ $\frac{1}{2}\ -\frac{1}{2} \to -\frac{1}{2}\ -\frac{1}{2}$	3 *lines corresponding to* $x + iy$ *of* (5)
‖	$0 \to 0$	$\frac{1}{2}\ \frac{1}{2} \to \frac{1}{2}\ \frac{1}{2}$ $\frac{1}{2}\ \frac{1}{2} \to -\frac{1}{2}\ -\frac{1}{2}$ $-\frac{1}{2}\ -\frac{1}{2} \to \frac{1}{2}\ \frac{1}{2}$ $-\frac{1}{2}\ -\frac{1}{2} \to -\frac{1}{2}\ -\frac{1}{2}$	4 *lines corresponding to* z *of* (5)
⊥	$-1 \to 0$	$-\frac{1}{2}\ \frac{1}{2} \to \frac{1}{2}\ \frac{1}{2}$ $-\frac{1}{2}\ \frac{1}{2} \to -\frac{1}{2}\ -\frac{1}{2}$ $-\frac{3}{2}\ -\frac{1}{2} \to -\frac{1}{2}\ -\frac{1}{2}$	3 *lines corresponding to* $x - iy$ *of* (5)

The possible transitions for m_l are $1 \rightarrow 0$, $0 \rightarrow 0$, $-1 \rightarrow 0$, corresponding to circular polarisation in one sense in the xy plane, linear polarisation along z, and circular polarisation in the opposite sense in the xy plane; these when viewed across the field, are $\perp$, $||$, $\perp$ components.

The possible transitions are shown above (p. 207) by arrows, since no Δm can exceed 1.

These are the 10 Zeeman lines of the figure of § 39 for the sodium D-doublet; the first three and the last three make the six r components, and the middle four the four p components of the figure.

To find the intensity of the line

$$m = \frac{1}{2},\ m_s = \frac{1}{2} \rightarrow m = \frac{1}{2},\ m_s = \frac{1}{2},$$

which is one of the four p lines, we carry out the transformation

$$|z'|^2 = S\,|z|^2 S^{-1}$$
$$= S\,|z|^2 \tilde{S}^*$$

so that

$$|z'|^2 (\tfrac{1}{2}\,\tfrac{1}{2}) = S\,(\tfrac{1}{2}\,\tfrac{1}{2})\,|z^2|\,(\tfrac{1}{2}\,\tfrac{1}{2})\,\tilde{S}^*\,(\tfrac{1}{2}\,\tfrac{1}{2})$$
$$= |S\,(\tfrac{1}{2}\,\tfrac{1}{2})|^2 . |z^2|\,(\tfrac{1}{2}\,\tfrac{1}{2}),$$

where the quantum numbers refer to m_s.

From (5),

$$|z|^2 (\tfrac{1}{2}\,\tfrac{1}{2}) = |B|^2 (l^2 - m_l^2) = |B|^2 (1^2 - 0^2) = |B|^2,$$

and from (3), $|S\,(\tfrac{1}{2}\,\tfrac{1}{2})|^2 = \tfrac{1}{4}\,|C|^2 (\mu + \tfrac{1}{2}\lambda + \Delta)^2$,

where

$$|C|^2 = \frac{2}{(\mu + \frac{1}{2}\lambda + \Delta)\,\Delta}$$

and

$$\Delta = \sqrt{\mu^2 + \mu\lambda + \tfrac{9}{4}\lambda^2},$$

writing $m = \frac{1}{2}$ and $l = 1$.

$$\therefore\ |z'|^2 (\tfrac{1}{2}\,\tfrac{1}{2}) = \frac{|B|^2}{2}\,\frac{\mu + \frac{1}{2}\lambda + \Delta}{\Delta}$$
$$= \frac{|B|^2}{2}\left\{1 + \frac{\mu + \frac{1}{2}\lambda}{\sqrt{\mu^2 + \mu\lambda + \frac{9}{4}\lambda^2}}\right\}.$$

Finally for the 10 lines, Heisenberg and Jordan found

$$\left.\begin{aligned}
|z'|^2\left(\tfrac{1}{2}\,\tfrac{1}{2},\,\tfrac{1}{2}\,\tfrac{1}{2}\right) &= \tfrac{1}{2}\,|B|^2\left(1+\frac{\mu+\frac{1}{2}\lambda}{\Delta}\right)\\
|z'|^2\left(-\tfrac{1}{2}-\tfrac{1}{2},\,\tfrac{1}{2}\,\tfrac{1}{2}\right) &= \tfrac{1}{2}\,|B|^2\left(1-\frac{\mu+\frac{1}{2}\lambda}{\Delta}\right)\\
|z'|^2\left(\tfrac{1}{2}\,\tfrac{1}{2},\,-\tfrac{1}{2}-\tfrac{1}{2}\right) &= \tfrac{1}{2}\,|B|^2\left(1+\frac{-\mu+\frac{1}{2}\lambda}{\Delta'}\right)\\
|z'|^2\left(-\tfrac{1}{2}-\tfrac{1}{2},\,-\tfrac{1}{2}-\tfrac{1}{2}\right) &= \tfrac{1}{2}\,|B|^2\left(1-\frac{-\mu+\frac{1}{2}\lambda}{\Delta'}\right)
\end{aligned}\right\}$$

$$\left.\begin{aligned}
|x'-iy'|^2\left(-\tfrac{1}{2}\,\tfrac{1}{2},\,\tfrac{1}{2}\,\tfrac{1}{2}\right) &= |B|^2\left(1+\frac{\mu-\frac{1}{2}\lambda}{\Delta'}\right)\\
|x'-iy'|^2\left(-\tfrac{1}{2}\,\tfrac{1}{2},\,-\tfrac{1}{2}-\tfrac{1}{2}\right) &= |B|^2\left(1-\frac{\mu-\frac{1}{2}\lambda}{\Delta'}\right)\\
|x'-iy'|^2\left(-\tfrac{3}{2}-\tfrac{1}{2},\,-\tfrac{1}{2}-\tfrac{1}{2}\right) &= |B|^2.2
\end{aligned}\right\}$$

$$\left.\begin{aligned}
|x'+iy'|^2\left(\tfrac{3}{2}\,\tfrac{1}{2},\,\tfrac{1}{2}\,\tfrac{1}{2}\right) &= |B|^2.2\\
|x'+iy'|^2\left(\tfrac{1}{2}-\tfrac{1}{2},\,\tfrac{1}{2}\,\tfrac{1}{2}\right) &= |B|^2\left(1-\frac{\mu+\frac{1}{2}\lambda}{\Delta}\right)\\
|x'+iy'|^2\left(\tfrac{1}{2}-\tfrac{1}{2},\,-\tfrac{1}{2}-\tfrac{1}{2}\right) &= |B|^2\left(1+\frac{\mu+\frac{1}{2}\lambda}{\Delta}\right)
\end{aligned}\right\}$$

where $\Delta = \sqrt{\mu^2+\lambda\mu+\frac{9}{4}\lambda^2}$,

and $\Delta' = \sqrt{\mu^2-\lambda\mu+\frac{9}{4}\lambda^2}$,

and the notation (m, m_s, m', m_s') means a transition in which $m \rightarrow m'$ and $m_s \rightarrow m_s'$, as indicated in the table of p. 207.

126. *Deduction of the Landé g and γ formulae.*

For a weak field $\mu \ll \lambda$.

If μ were zero, then $\overline{H}_1 = 0$ and

$$\overline{H}_2 = \frac{e^2Z}{2{m_0}^2c^2}\left(\overline{\frac{1}{r^3}}\right)\mathbf{L}.\mathbf{S} = \left(\frac{2\pi}{h}\right)^2\lambda\,\mathbf{L}.\mathbf{S}.$$

If $\mathbf{J}$ is the total angular momentum vector for the atom,

$$\mathbf{J} = \mathbf{L}+\mathbf{S}$$

and
$$\begin{aligned}\mathbf{J}^2 &= \mathbf{L}^2+\mathbf{S}^2+\mathbf{L}.\mathbf{S}+\mathbf{S}.\mathbf{L}\\ &= \mathbf{L}^2+\mathbf{S}^2+2\mathbf{L}.\mathbf{S},\end{aligned}$$

since $\mathbf{L}$, $\mathbf{S}$ commute.

So $$\mathbf{L}^2 = \mathbf{J}^2+\mathbf{S}^2-2\mathbf{J}.\mathbf{S} \quad\text{.................}(1).$$

$$\therefore \quad 2\mathbf{L}.\mathbf{S} = \mathbf{J}^2 - \mathbf{L}^2 - \mathbf{S}^2$$
$$= \left(\frac{h}{2\pi}\right)^2 [j(j+1) - l(l+1) - s(s+1)].$$
$$\therefore \quad \bar{H}_2 = \tfrac{1}{2}\lambda\,[j(j+1) - l(l+1) - s(s+1)].$$

The actual system, if μ is small, is this one slightly perturbed. In the undisturbed system the atom precesses about **J**. The energy values of the perturbed system are given by the mean value of H_1 over the undisturbed path. If **L**, **S** are resolved into one component parallel to **J** and one perpendicular to **J**, the two latter will disappear, on account of the precession, on taking mean values and only the former give a contribution to H_1. Taking over into the quantum theory these considerations borrowed from the classical mechanics, we have, since all the magnitudes concerned commute, the components along **J** of **L** and **S** are

$$\frac{\mathbf{J}.\mathbf{L}}{\mathbf{J}^2}\mathbf{J} \text{ and } \frac{\mathbf{J}.\mathbf{S}}{\mathbf{J}^2}\mathbf{J}.$$

The corresponding magnetic moments are

$$\frac{\mathbf{J}.\mathbf{L}}{\mathbf{J}^2}\mathbf{J}\frac{e}{2m_0c}, \quad \frac{\mathbf{J}.\mathbf{S}}{\mathbf{J}^2}\mathbf{J}\frac{e}{m_0c}.$$

Thus $$\bar{H}_1 = \frac{e}{2m_0c}\mathbf{H}.\mathbf{J}\left\{\frac{\mathbf{J}.\mathbf{L}}{\mathbf{J}^2} + \frac{2\mathbf{J}.\mathbf{S}}{\mathbf{J}^2}\right\}$$
$$= \frac{e}{2m_0c}\mathbf{H}.\mathbf{J}\left\{1 + \frac{\mathbf{J}.\mathbf{S}}{\mathbf{J}^2}\right\} = \mu m\left\{1 + \frac{\mathbf{J}^2 - \mathbf{L}^2 + \mathbf{S}^2}{2\mathbf{J}^2}\right\},$$

using (1), $$= \mu m\left\{1 + \frac{j(j+1) - l(l+1) + s(s+1)}{2j(j+1)}\right\}.$$

So that in general for $\mu \ll \lambda$,

$$\bar{H}_1 + \bar{H}_2 = \mu m\left\{1 + \frac{j(j+1) - l(l+1) + s(s+1)}{2j(j+1)}\right\}$$
$$+ \tfrac{1}{2}\lambda\{j(j+1) - l(l+1) - s(s+1)\},$$

agreeing with the Landé formulae for the g and γ values.

127. *Calculation of the fine structure due to relativity and spin.*

We have to calculate

$$\lambda = \frac{Ze^2}{2m_0{}^2c^2}\left(\frac{h}{2\pi}\right)^2\overline{\left(\frac{1}{r^3}\right)}$$

and $$\bar{H}_3 = -\frac{1}{2m_0c^2}\left\{E_0{}^2 + 2Ze^2E_0\overline{\left(\frac{1}{r}\right)} + Z^2e^4\overline{\left(\frac{1}{r^2}\right)}\right\},$$

or we have to find $\overline{\left(\frac{1}{r}\right)}, \overline{\left(\frac{1}{r^2}\right)}, \overline{\left(\frac{1}{r^3}\right)}$.

We first use the theorem that for a Coulomb field $\bar{T} = -\frac{1}{2}\bar{V}$, where T, V are the kinetic and potential energies. For if V is homogeneous of degree n in the coordinates q,

$$nV = \Sigma q\frac{\partial V}{\partial q}.$$

But $$\frac{d}{dt}\Sigma pq = \Sigma(\dot{p}q + p\dot{q}) = \Sigma\left(-q\frac{\partial H}{\partial q} + p\frac{\partial H}{\partial p}\right),$$

and if $$H = T + V = \frac{1}{2}\Sigma\frac{p_k{}^2}{m_k} + V(q),$$

$$\frac{d}{dt}\Sigma pq = \Sigma\left(-q\frac{\partial V}{\partial q} + p\frac{p}{m}\right)$$
$$= -nV + 2T.$$

Taking the time mean, $2\bar{T} = n\bar{V}$. For the Coulomb field $n = -1$ so that $2\bar{T} = -\bar{V}$.

$$\therefore\ -\overline{\frac{Ze^2}{r}} = 2\bar{T} = 2E_0 \text{ and } E_0 = -\frac{RhcZ^2}{n^2}.$$

$$\therefore\ \overline{\left(\frac{1}{r}\right)} = \frac{2RhcZ}{n^2e^2} \quad\quad \text{..................(1).}$$

Also the Pauli-Dirac Hamiltonian is (§ 83)

$$H_0 = \frac{1}{2m_0}\left\{p_r^2 + \frac{1}{r^2}\left(p_\theta^2 - \frac{h^2}{16\pi^2}\right)\right\} - \frac{Ze^2}{r},$$

so that $$\dot{p}_r = -\frac{\partial H_0}{\partial r} = \frac{1}{m_0 r^3}\left(p_\theta^2 - \frac{h^2}{16\pi^2}\right) - \frac{Ze^2}{r^2}.$$

Therefore by a time mean,

$$0 = \overline{\frac{e^2 Z}{r^2}} - \overline{\frac{1}{m_0 r^3}}\left(p_\theta^2 - \frac{h^2}{16\pi^2}\right) \quad \text{............(2).}$$

Also $$m_0 r^2 \dot{\theta} = p_\theta.$$

$$\therefore \quad \overline{\frac{p_\theta}{m_0 r^2}} = \overline{\dot{\theta}} \quad \text{........................(3).}$$

Introduce the angle variable w corresponding to J $(= nh)$.

Then $$\overline{\dot{\theta}} = 2\pi\dot{w} = 2\pi\frac{\partial H_0}{\partial J} = \frac{2\pi}{h}\frac{\partial E_0}{\partial n} = \frac{4\pi RcZ^2}{n^3}.$$

Hence since[1] $p_\theta = \frac{h}{2\pi}(l + \frac{1}{2})$, we have from (3)

$$\left(\overline{\frac{1}{r^2}}\right) = \frac{2\pi m_0}{h(l + \frac{1}{2})}\frac{4\pi RcZ^2}{n^3} = \frac{8\pi^2 m_0 RcZ^2}{h(l + \frac{1}{2})\,n^3} \quad \text{......(4).}$$

But from (2),

$$\left(\overline{\frac{1}{r^3}}\right) = \frac{m_0 e^2 Z\left(\overline{\frac{1}{r^2}}\right)}{\{(l + \frac{1}{2})^2 - \frac{1}{4}\}\frac{h^2}{4\pi^2}}$$

$$= \frac{32 m_0^2 e^2 \pi^4 RcZ^3}{l(l + \frac{1}{2})(l + 1)\,n^3 h^3}, \text{ using (4),...(5).}$$

$$\therefore \quad \lambda = \frac{Ze^2}{2m_0^2 c^2}\left(\frac{h}{2\pi}\right)^2\left(\overline{\frac{1}{r^3}}\right)$$

$$= \frac{4\pi^2 Z^4 e^4 R}{chn^3 l(l + \frac{1}{2})(l + 1)}$$

$$= \frac{C\alpha^2 Z^4}{n^3 l(l + \frac{1}{2})(l + 1)},$$

[1] From Pauli's paper on hydrogen, *l.c.* p. 127.

where $C = Rch$, R being the Rydberg constant,

$$\frac{2\pi^2 m_0 e^4}{ch^3} \equiv 109677, \text{ and } \alpha = \frac{2\pi e^2}{hc},$$

the fine structure constant (cf. § 35).

$$\therefore \quad \bar{H}_2 = \frac{\lambda}{2}\{j(j+1) - l(l+1) - s(s+1)\}$$

$$= \frac{C\alpha^2 Z^4}{2n^3} \cdot \frac{j(j+1) - l(l+1) - s(s+1)}{l(l+\frac{1}{2})(l+1)}.$$

Again

$$\bar{H}_3 = -\frac{1}{2m_0 c^2}\left\{E_0^2 + 2Ze^2 E_0 \overline{\left(\frac{1}{r}\right)} + Z^2 e^4 \overline{\left(\frac{1}{r^2}\right)}\right\}$$

$$= -\frac{1}{2m_0 c^2}\left\{\left(\frac{RhcZ^2}{n^2}\right)^2 - 2Ze^2 . \frac{RhcZ^2}{n^2} \cdot \frac{2RhcZ}{n^2 e^2} + Z^2 e^4 \frac{8\pi^2 m_0 RcZ^2}{h(l+\frac{1}{2})n^3}\right\}$$

$$= -\frac{1}{2m_0 c^2}\left\{-3\left(\frac{RchZ^2}{n^2}\right)^2 + \frac{4h^2 R^2 c^2 Z^4}{n^3(l+\frac{1}{2})}\right\}$$

$$= \frac{C\alpha^2 Z^4}{n^3}\left(\frac{3}{4n} - \frac{1}{l+\frac{1}{2}}\right).$$

Hence the energy of an (n, j, l) state is

$$\frac{CZ^2}{n^2} + \frac{C\alpha^2 Z^4}{n^3}\left\{\frac{j(j+1) - l(l+1) - s(s+1)}{2l(l+\frac{1}{2})(l+1)}\right\} + \frac{C\alpha^2 Z^4}{n^3}\left(\frac{3}{4n} - \frac{1}{l+\frac{1}{2}}\right).$$

This result is that used in § 35 to clear up the riddle of the X-ray doublets.

128. *The relation of the spinning electron to Schrödinger's wave mechanics.*

Pauli[1] has recently done some work on the application of quantum mechanics to the spinning electron, which is now being extended by Jordan so as to include relativity.

[1] W. PAULI (to appear shortly).

Pauli's theory consists in finding matrices or operators to represent the components of the spin angular momentum, s_x, s_y, s_z, which must satisfy the usual relations $[s_x s_y] = s_z$, etc.

The matrices

$$s_x = \frac{h}{4\pi}\begin{pmatrix} 0 & i \\ -i & 0 \end{pmatrix}, \quad s_y = \frac{h}{4\pi}\begin{pmatrix} 0 & 1 \\ 1 & 0 \end{pmatrix}, \quad s_z = \frac{h}{4\pi}\begin{pmatrix} 1 & 0 \\ 0 & 1 \end{pmatrix},$$

for instance satisfy these relations.

If now the wave function ψ is regarded as having two components ψ_1, ψ_2, one can give a meaning to any function of s_x, s_y, s_z operating upon ψ. For example, to evaluate $s_x\psi$, we have

$$(s_x\psi)_1 = s_x(11)\,\psi_1 + s_x(12)\,\psi_2 = 0.\psi_1 + \frac{ih}{4\pi}\psi_2,$$

$$(s_x\psi)_2 = s_x(21)\,\psi_1 + s_x(22)\,\psi_2 = -\frac{ih}{4\pi}\psi_1 + 0.\psi_2.$$

In this way the Schrödinger equation $(H - E)\,\psi = 0$, where H is an operator including s_x, s_y, s_z as well as q and $\frac{\partial}{\partial q}$, can be interpreted as two simultaneous differential equations in ψ_1 and ψ_2, which can be solved. In this way[1] the spin can be taken into the Schrödinger scheme.

Darwin[2] has also worked out the motion of a spinning electron moving in a central orbit in a magnetic field by the use of wave mechanics and spherical harmonics. The wave equation is generalised by supposing the electron-magnet to be represented by polarised waves. Jordan[3] has recently shown that the polarisation properties of a light quantum are formally equivalent to Pauli's theory of the electron-magnet.

[1] PAULI's paper has since appeared in the Zs. f. Phys. **43**, p. 601, 1927.

[2] C. G. DARWIN, Proc. Roy. Soc. A. **115**, p. 1, 1927 and **116**, p. 227, 1927.

[3] P. JORDAN, Zs. f. Phys. **44**, p. 292, 1927.

CHAPTER XXVI

HEISENBERG'S RESONANCE THEORY OF THE ORTHO AND PARA HELIUM SPECTRA

129. *The spectrum of neutral helium.*

It is well known that the spectral terms of helium can be divided into two sets such that no term of the one will combine with a term of the other to produce a spectral line. Both sets are approximately like hydrogen terms.

One set by its transitions gives the 'para helium' lines and consists of singulet terms and to it belongs the 'normal state' $1s$; the other set gives the 'ortho helium' lines and (apart from the singulet terms) consists of very narrow doublets (theory suggests that these are probably triplets). The energies of the ortho terms are slightly higher than those of the corresponding para terms.

On account of these two sets of lines, helium had been thought to be a mixture of two gases, para and ortho helium.

Many attempts[1] were made to give the theory of these spectra on the earlier quantum theory, both for normal and excited helium, but all failed to give results even in tolerable agreement with experiment.

The obvious failure of the classical mechanics and the correspondence principle to solve the problem of a nucleus with two outer electrons was one of the factors which compelled Heisenberg to seek for a new quantum

[1] N. Bohr, Phil. Mag. **26**, p. 476, **1913**. A. Landé, Phys. Zs. **20**, p. 228, **1919**. H. A. Kramers, Zs. f. Phys. **13**, p. 312, **1923**. J. H. van Vleck, Phys. Rev. **21**, p. 372, **1923**. M. Born and W. Heisenberg, Zs. f. Phys. **26**, p. 216, **1924**.

mechanics of discontinuous processes to replace the older theory.

Heisenberg has in a recent paper[1] given the clue to the ortho-para separation, which consists in a 'resonance' phenomenon, and in a second paper[2] has found numerical results which give in a rough degree a quantitative account of the He spectrum. The series used in the perturbation theory converge sufficiently rapidly for the first approximation used to give the highly excited d, b terms reasonably well; but the p terms are given less exactly and the s terms hardly at all. For more exact calculations, finer methods of approximation will have to be found which will give rapidly converging series, such as were used by Kramers in the earlier quantum theory of the helium atom.

130. *Heisenberg's resonance theory of the ortho and para helium terms.*

The simplest many-body problem is a system of two coupled oscillators. This problem possesses all the characteristic properties of the quantum theory many-body problem and results can be obtained from it which clear up the theory of the spectra.

It is characteristic of atomic systems that the electrons which compose them are equal and are subject on the whole to the same forces; for He each electron is subject to the same action on the whole, and the coupling is due to the Coulomb field between them.

Heisenberg therefore considers two exactly equal oscillators of frequency ν whose total Hamiltonian H

$$= \frac{1}{2m} p_1^2 + \frac{m}{2} (2\pi\nu)^2 q_1^2 + \frac{1}{2m} p_2^2 + \frac{m}{2} (2\pi\nu)^2 q_2^2 + \lambda q_1 q_2,$$

where the term $\lambda q_1 q_2$ is the interaction or coupling energy of the oscillators. When the system is reduced to its

[1] W. HEISENBERG, Zs. f. Phys. **38**, p. 411, **1926**.
[2] W. HEISENBERG, Zs. f. Phys. **39**, p. 499, **1926**.

principal coordinates Q_1, Q_2 by the usual methods of dynamics, it is found that

$$Q_1 = \frac{1}{\sqrt{2}}(q_1 + q_2) \text{ and } Q_2 = \frac{1}{\sqrt{2}}(q_1 - q_2),$$

and that if P_1, P_2 are the corresponding momenta, H becomes

$$\frac{1}{2m}P_1^2 + \frac{m}{2}(2\pi\nu_1)^2 Q_1^2 + \frac{1}{2m}P_2^2 + \frac{m}{2}(2\pi\nu_2)^2 Q_2^2,$$

where $$\nu_1^2 = \nu^2 + \frac{\lambda}{4\pi^2}, \quad \nu_2^2 = \nu^2 - \frac{\lambda}{4\pi^2}.$$

H is now the sum of two oscillator energies which represent the two principal vibrations of the coupled systems.

If Q_1 is excited, then $Q_2 = 0$ and $q_1 = q_2$, so that the two particles vibrate in the same phase; if Q_2 is excited, $Q_1 = 0$ and $q_1 = -q_2$, so that the particles vibrate in opposite phase.

The energies of the stationary states of the whole system are given by

$$H_{n_1 n_2} = \nu_1 h(n_1 + \tfrac{1}{2}) + \nu_2 h(n_2 + \tfrac{1}{2}), \quad (\S\ 57).$$

The correspondence principle is now used to determine the possible transitions between the different terms $H_{n_1 n_2}$. (The notation 20, for example, of the figure means a term for which $n_1 = 2$, $n_2 = 0$.)

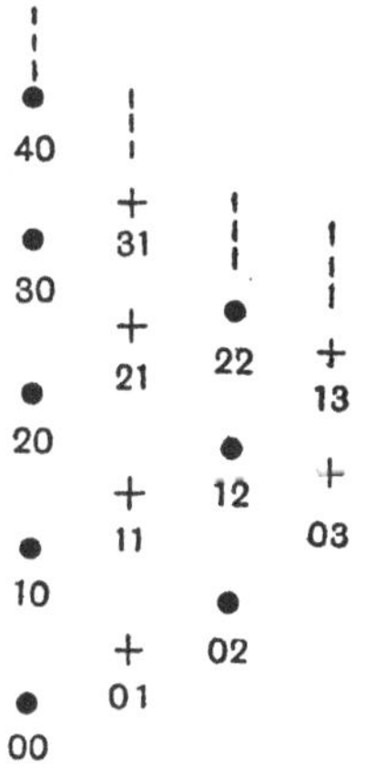

If for simplicity we regard the particles as point charges, the dipole moment is (except for a constant factor) $q_1 + q_2$. Thus only Q_1 has an electric moment, and since Q_1 is of the form $e^{2\pi i\nu_1 t}$, that is a Fourier series $\Sigma_\tau e^{2\pi i\tau\nu_1 t}$ with only the term $\tau = 1$, transitions can occur only where the corresponding quantum number n_1 changes by unity. Thus to

40 31 30 22 21 13 20 12 11 03 10 02 01 00

this approximation transitions can only occur between successive terms of the figure which lie in the same vertical line.

But the dipole moment only gives the radiation to a first approximation; quadripoles and higher poles (arising from collision processes) give rise to terms in the radiation of a lower order of magnitude. These terms are given by symmetric functions of the second, third and higher orders in q_1, q_2 and their differential coefficients. A term of such a symmetric function, say of only q_1, q_2, would be

$$q_1^r q_2^s + q_2^r q_1^s$$

or $$\left(\frac{Q_1+Q_2}{\sqrt{2}}\right)^r \left(\frac{Q_1-Q_2}{\sqrt{2}}\right)^s + \left(\frac{Q_1-Q_2}{\sqrt{2}}\right)^r \left(\frac{Q_1+Q_2}{\sqrt{2}}\right)^s$$

and would only contain *even* powers of Q_2.

Since Q_2 is of the form $e^{2\pi i\nu_2 t}$, such a symmetric function can only contain the ν_2 frequency in terms of the form $e^{(2\pi i\nu_2 t)2\tau}$, where τ is an integer, so that the corresponding quantum number n_2 can only change by 2τ, that is by an *even* number.

Therefore transitions can only occur between terms within the system ● or between terms within the system +, but not between a term of the system ● and a term of the system +.

Thus this system of equal coupled oscillators has the fundamental property of *non-combination* of two sets of terms possessed by the ortho and para terms of helium.

Having discovered this clue to the solution of the helium problem, Heisenberg points out that *all* these terms are energy levels, and that a sound scheme of quantum mechanics should be able to show that the *amplitude* of the transition from a ● term to a + term is *zero*. He then proceeds to prove this in the following manner.

131. *Resonance in quantum mechanics.*

He considers two exactly similar systems a, b which are coupled by an interaction energy H symmetrical in the coordinates of both systems. The energies of their stationary states (alone) are H_m^a, H_n^b say. The total energy, when uncoupled, is $H_{mn} \equiv H_m^a + H_n^b$ corresponding to a stationary state (mn), whose quantum numbers are m, n. Since the systems are identical, $H_m^a = H_m^b$ and $H_n^a = H_n^b$, so that $H_{mn} = H_{nm}$.

Further $H_m^a - H_n^a = H_m^b - H_n^b$, which means that the system a in passing from the state m to the state n is emitting just the frequency to be absorbed by b to cause it to pass from the state n to the state m. This is *resonance* as understood in quantum mechanics. Resonance can only occur if the systems are in *different energy states*, one in the state m, the other in the state n, a consideration which does not enter into the classical theory of resonance. Resonance in quantum mechanics is essentially different from resonance in the classical theory; it is this difference, which cannot be bridged by correspondence principle methods, which accounts for the failure of the earlier attempts to solve the helium problem.

The above system (a and b uncoupled) is degenerate, for since $H_{mn} = H_{nm}$, every energy level is duplicated, except the one for which $n = m$.

The coupling action of energy H however removes this degeneracy. Treating this as a perturbation and using the theory of § 63, we first find the mean value $\bar{H}$ of H for the undisturbed system. This will however not be a diagonal matrix but will also contain constant terms corresponding to transitions in which the systems a, b pass from a state m, n to a state n, m in the undisturbed system without energy emission (since $H_{mn} = H_{nm}$).

Thus the matrix for $\bar{H}$ contains only diagonal terms $\bar{H}\,(mn, mn)$ and also terms $\bar{H}\,(nm, mn)$. The other terms are zero.

Proceeding as in § 63, we make the transformation

$$S\bar{\bar{H}}S^{-1} = E, \text{ so that } E \text{ is a diagonal matrix.}$$

$$\therefore \; ES = S\bar{\bar{H}}.$$

Then

$$\Sigma_{kl} E\,(nm, kl)\, S\,(kl, nm) = \Sigma_{kl} S\,(nm, kl)\, \bar{\bar{H}}\,(kl, nm)$$

or
$$E\,(nm, nm)\, S\,(nm, nm) = S\,(nm, nm)\, \bar{\bar{H}}\,(nm, nm) + S\,(nm, mn)\, \bar{\bar{H}}\,(mn, nm);$$

or writing E_{nm} for $E\,(nm, nm)$, we have

$$E_{nm}\, S\,(nm, nm) = S\,(nm, nm)\, \bar{\bar{H}}\,(nm, nm) + S\,(nm, mn)\, \bar{\bar{H}}\,(mn, nm) \ldots\ldots (1).$$

Again

$$\Sigma_{kl} E\,(nm, kl)\, S\,(kl, mn) = \Sigma_{kl} S\,(nm, kl)\, \bar{\bar{H}}\,(kl, mn),$$

whence

$$E_{nm}\, S\,(nm, mn) = S\,(nm, nm)\, \bar{\bar{H}}\,(nm, mn) + S\,(nm, mn)\, \bar{\bar{H}}\,(mn, mn) \ldots\ldots (2).$$

On account of the symmetry of $\bar{\bar{H}}$ in the coordinates of a and b,

$$\bar{\bar{H}}\,(nm, nm) = \bar{\bar{H}}\,(mn, mn) = \lambda,$$
$$\bar{\bar{H}}\,(nm, mn) = \bar{\bar{H}}\,(mn, nm) = \mu, \text{ suppose,}$$

so that (1) and (2) become

$$\left.\begin{aligned}(E_{nm} - \lambda)\, S\,(nm, nm) &= \mu\, S\,(nm, mn)\\ (E_{nm} - \lambda)\, S\,(nm, mn) &= \mu\, S\,(nm, nm)\end{aligned}\right\} \ldots\ldots (3).$$

$$\therefore \; (E_{nm} - \lambda)^2 = \mu^2.$$

So

$$\left.\begin{aligned}(E_{mn} - \lambda)\, S\,(mn, mn) &= \mu\, S\,(mn, nm)\\ (E_{mn} - \lambda)\, S\,(mn, nm) &= \mu\, S\,(mn, mn)\end{aligned}\right\} \ldots\ldots (4).$$

$$\therefore \; (E_{mn} - \lambda)^2 = \mu^2.$$

Thus E_{mn}, E_{nm} are the roots of the quadratic

$$(E - \lambda)^2 = \mu^2, \text{ in } E.$$

Therefore we write $E_{nm} = \lambda + \mu$ and $E_{mn} = \lambda - \mu$.

Therefore from (3) and (4),

$$\left.\begin{aligned}S\,(nm, nm) &= S\,(nm, mn) = \theta\\ S\,(mn, mn) &= -S\,(mn, nm) = \phi\end{aligned}\right\}, \text{ suppose, } \ldots (5).$$

Normalising S by $S\tilde{S}^* = 1$, we have

$$\left.\begin{aligned}\Sigma_{kl} S(nm, kl)\, \tilde{S}^*(kl, nm) &= 1\\ \Sigma_{kl} S(mn, kl)\, \tilde{S}^*(kl, mn) &= 1\end{aligned}\right\},$$

or

$$\left.\begin{aligned}\Sigma_{kl} S(nm, kl)\, S^*(nm, kl) &= 1\\ \Sigma_{kl} S(mn, kl)\, S^*(mn, kl) &= 1\end{aligned}\right\},$$

or

$$\left.\begin{aligned}S(nm, mn)\, S^*(nm, mn) + S(nm, nm)\, S^*(nm, nm) &= 1\\ S(mn, mn)\, S^*(mn, mn) + S(mn, nm)\, S^*(mn, nm) &= 1\end{aligned}\right\}.$$

$$\therefore \quad \left.\begin{aligned}2\theta\theta^* &= 1\\ 2\phi\phi^* &= 1\end{aligned}\right\}.$$

$$\therefore \quad |\theta|^2 = |\phi|^2 = \tfrac{1}{2},$$

and taking the phase into the time factor as usual, we have

$$\theta = \frac{1}{\sqrt{2}}, \; \phi = \frac{1}{\sqrt{2}}.$$

Thus the transformation matrix S is given by

$$\left.\begin{aligned}S(nm, nm) = \quad S(nm, mn) &= \frac{1}{\sqrt{2}}\\ S(mn, mn) = -S(mn, nm) &= \frac{1}{\sqrt{2}}\end{aligned}\right\} \quad \ldots\ldots\ldots (6),$$

and

$$\left.\begin{aligned}E_{nm} &= \bar{H}(nm, nm) + \bar{H}(nm, mn)\\ E_{mn} &= \bar{H}(nm, nm) - \bar{H}(nm, mn)\end{aligned}\right\} \; (7).$$

Thus the coupling makes E_{nm} different from E_{mn} and the separation

$$E_{nm} - E_{mn} = 2\bar{H}(nm, mn) \ldots (8).$$

The figure shows the term spectrum for the coupled systems where for example 42 denotes a term for which the energy is E_{42}, $(n = 4, m = 2)$.

It will now be shown that no term of the set ● can combine with any term of the set +.

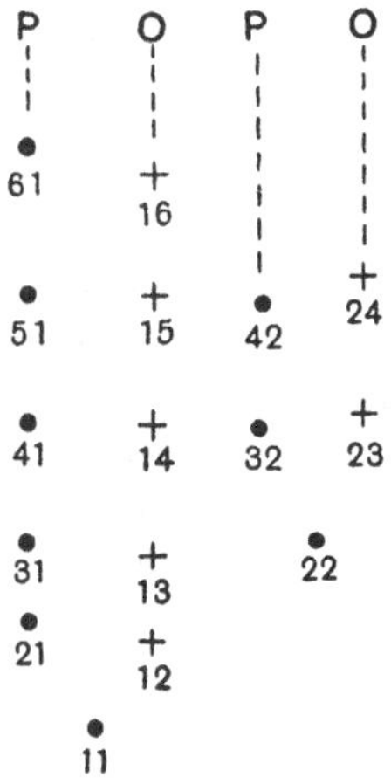

This is effected by using the new mechanics to calculate the amplitude of the radiation due to a transition from a state ● to a state + and proving the amplitude to be zero.

132. *Calculation of the amplitudes.*

Let the radiation be given by a function f of the co-ordinates p, q of the uncoupled system, so that $f_{nm} = f_{mn}$.

For the coupled system the radiation is given by $f' = SfS^{-1}$, (§ 60) so that $f' = Sf\tilde{S}^*$, since $S\tilde{S}^* = 1$.

$$\therefore\ f'(n_1m, mn_2) = \sum_{klrs} S(n_1m, kl) f(kl, rs) \tilde{S}^*(rs, mn_2)$$
$$= \Sigma S(n_1m, kl) f(kl, rs) S^*(mn_2, rs)$$
$$= S(n_1m, n_1m)\{f(n_1m, mn_2) S^*(mn_2, mn_2) + f(n_1m, n_2m) S^*(mn_2, n_2m)\}$$
$$+ S(n_1m, mn_1)\{f(mn_1, mn_2) S^*(mn_2, mn_2) + f(mn_1, n_2m) S^*(mn_2, n_2m)\},$$

using the only possible values of k, l, r, s for which the matrix components of S or S^* are not zero.

$$\therefore\ f'(n_1m, mn_2) = \tfrac{1}{2}\{f(n_1m, mn_2) - f(n_1m, n_2m) + f(mn_1, mn_2) - f(mn_1, n_2m)\},$$ using (6), § 131.

But since $f_{mn} = f_{nm}$,

$$f(n_1m, mn_2) = f(mn_1, n_2m)$$

and
$$f(n_1m, n_2m) = f(mn_1, mn_2),$$

so that
$$f'(n_1m, mn_2) = 0 \qquad \text{(1)}.$$

Thus the amplitude of a transition from the state (n_1m) to the state (mn_2) is zero; in other words the corresponding line is not observed. Thus for example no transitions can occur between any one of the states 21, 31, 41, ... and any one of the states 12, 13, 14, ...

In the same way,

$$f'(nm_1, m_2n) = 0 \qquad \text{(2)},$$

$$\left.\begin{aligned} f'(n_1m, n_2m) &= f(n_1m, n_2m) + f(n_1m, mn_2) \\ f'(nm_1, nm_2) &= f(nm_1, nm_2) - f(nm_1, m_2n) \end{aligned}\right\} \quad \text{(3)},$$

and also $f'(n_1 m, mm) = \sqrt{2} f(n_1 m, mm)$ ....(4),

$f'(mm_1, mm) = 0$ ....(5).

The results (1), (2) and (5) show that no transitions can occur between any one of the ● terms and any one of the + terms of the figure of § 131.

The line intensities within each term system ● or + are on account of (3) to a first approximation the same as those of the corresponding terms of the uncoupled system, since to a first approximation $f(n_1 m, mn_2) = 0$, from (1).

But from (4)

$$|f'(n_1 m, mm)|^2 = 2\,|f(n_1 m, mm)|^2,$$

so that the intensity (measured by $|f|^2$, § 67) of a line such as that due to the terms $41 \rightarrow 11$ is twice as great as for the uncoupled system.

133. *Application to the helium spectrum.*

If in the helium problem, we assume the two electrons moving round the nucleus to be point charges (with no spin and consequent magnetic moment), the two term systems ●, + are the terms of *P*(*ara*) and *O*(*rtho*) helium. Transitions between the two sets of terms are not possible, as has been shown (equations 1, 2, 5, § 132).

Also the 1*s* term is in one only of the two systems and the theory shows (equations 3, 4, § 132), that it lies in the system (*P*) whose terms are of *higher* energy than the corresponding terms of the other system (*O*).

The theory shows that the intensities of the *P*- and *O*-lines are equal to those of the corresponding hydrogen lines, but that the *P*-lines which involve transitions to the term 1*s* are approximately twice as intense as for hydrogen.

The separation of a pair of corresponding *P*- and *O*-lines is $2\bar{H}(nm, mn)$, by equation (8), § 131, the $\bar{H}$ being the coupling energy due to the Coulomb forces between the electrons. The separation is due to *nm* changing to *mn* so that the two electrons are continually changing places and in this way maintain the *P-O* separation.

Thus the facts stated in § 130 are accounted for by this theory of §§ 131–2.

Wigner[1] has applied the Schrödinger volume density theory, § 103, to the case of a nucleus with three electrons, and Heisenberg[2] using the theory of groups has developed his resonance theory for the multiplets of higher atoms and the band spectra of molecules.

The theory of groups is destined to play an important part in the theory of the higher atoms, as is apparent from the work of Heisenberg, Wigner, and Hund[3].

[1] E. Wigner, Zs. f. Phys. **40**, pp. 492 and 883, **1926**.
[2] W. Heisenberg, Zs. f. Phys. **41**, p. 239, **1926**.
[3] F. Hund, Zs. f. Phys. **43**, p. 788, **1927**.

CHAPTER XXVII

THE EFFECT OF THE SPIN OF THE ELECTRONS UPON THE HELIUM SPECTRUM; THE CALCULATIONS OF HEISENBERG FOR HELIUM; SYMMETRIC AND ANTI-SYMMETRIC EIGENFUNCTIONS

134. *Application of Schrödinger's eigenfunctions to show the non-combination of O- and P-terms.*

For the original system a let $\psi_m{}^a$ be the eigenfunction corresponding to the energy $H_m{}^a$, and for b let $\psi_n{}^b$ be that corresponding to $H_n{}^b$. Suppose both eigenfunctions normalised.

In order to find a matrix element $x\,(n_1 n_2)$ of a coordinate x of the a system, we have

$$x\,(n_1 n_2) = \int x \,.\, \bar{\psi}_{n_1}{}^a \psi_{n_2}{}^a \,dq \quad (\S\,102) \,\ldots\ldots\ldots(1).$$

For the uncoupled pair a, b, the energy is $H_m{}^a + H_n{}^b$ and therefore the corresponding eigenfunction $\psi_m{}^a \psi_n{}^b$.

The eigenfunctions for the coupled system which correspond to E_{nm} and E_{mn} (§ 131) are derived from those of the uncoupled system by a linear transformation with the matrix S; to E_{nm} corresponds the eigenfunction

$$\frac{1}{\sqrt{2}} (\psi_m{}^a \psi_n{}^b + \psi_n{}^a \psi_m{}^b),$$

and to E_{mn} corresponds the eigenfunction

$$\frac{1}{\sqrt{2}} (\psi_m{}^a \psi_n{}^b - \psi_n{}^a \psi_m{}^b).$$

Hence by (1) above

$x\,(nm, mn)$

$$= \int x \frac{1}{\sqrt{2}} \overline{(\psi_m{}^a \psi_n{}^b + \psi_n{}^a \psi_m{}^b)} \frac{1}{\sqrt{2}} (\psi_m{}^a \psi_n{}^b - \psi_n{}^a \psi_m{}^b)\, dq.$$

If x is a magnitude, such as the radiation from the atom, which is symmetrical in a and b, the right-hand side is symmetrical in a, b and is unaffected by interchange of a, b. But clearly it changes sign if a, b are interchanged. Therefore it must be zero.

135. *The effect of the spin of the electrons*[1].

The result just found is our new undisturbed problem; we have now to find the perturbing effect of the spin of the electrons.

In the undisturbed problem there are two sets of terms, one set having eigenfunctions

$$\frac{1}{\sqrt{2}}(\psi_m{}^a\psi_n{}^b + \psi_n{}^a\psi_m{}^b),$$

which are *symmetric* (i.e. unaltered by interchange of a and b), the other set having eigenfunctions

$$\frac{1}{\sqrt{2}}(\psi_m{}^a\psi_n{}^b - \psi_n{}^a\psi_m{}^b),$$

which are *antisymmetric* (i.e. change sign if a and b are interchanged).

Also terms of the one set will not combine with terms of the other. Since the forces on the two electron-magnets are the same, they place themselves so as to be parallel or antiparallel, so that the usual quantum number m associated with their orientation in space must be $\pm\frac{1}{2}$ (for $|m| \leqslant s$, where $s = \frac{1}{2}$); we have as the possibilities for the two electrons in the undisturbed problem the table on p. 227.

In α and δ the quantising of m fixes m_1 and m_2, since each $= \frac{1}{2}m$. But it is not so for β and γ, since $m = 0$ leads to two possibilities and in this case there arises the resonance degeneracy of § 131.

[1] W. Heisenberg, Zs. f. Phys. **39**, p. 511, **1926**, and Zs. f. Phys. **40**, pp. 252–5, **1927**.

m_1	m_2	$m = m_1 + m_2$	
$\frac{1}{2}$	$\frac{1}{2}$	1	α
$\frac{1}{2}$	$-\frac{1}{2}$	0	β
$-\frac{1}{2}$	$\frac{1}{2}$	0	γ
$-\frac{1}{2}$	$-\frac{1}{2}$	-1	δ

The states α, δ correspond to states of the (mm) type of § 131 and the states β, γ to states of the (mn) type.

The effect of the perturbation is to produce four new states α', β', γ', δ' of which α', δ' are represented by symmetric eigenfunctions (since $m_1 = m_2$ for each of α, δ). But β', γ' are represented, one (β' suppose) by a symmetric eigenfunction, and the other, γ', by an antisymmetric one, as the theory of resonance degeneracy requires (§ 134).

Thus associated with each of the two original eigenfunctions

$$\frac{1}{\sqrt{2}}(\psi_m{}^a \psi_n{}^b + \psi_n{}^a \psi_m{}^b)$$

which is symmetric, and

$$\frac{1}{\sqrt{2}}(\psi_m{}^a \psi_n{}^b - \psi_n{}^a \psi_m{}^b)$$

which is antisymmetric, are four new eigenfunctions of which three are symmetric and one antisymmetric. The final eigenfunctions which represent the perturbed states are the products of each of the original ones into each of the four new ones arising from the spin of the electrons.

Thus the original symmetric eigenfunction (P-term) is resolved into three symmetric and one antisymmetric eigenfunctions, while the original antisymmetric eigenfunction (O-term) is resolved into three antisymmetric and one symmetric eigenfunctions.

From § 134, terms ● corresponding to antisymmetrical eigenfunctions cannot combine with terms + corresponding to symmetrical ones.

Hence the scheme of the figure of § 131 now becomes:

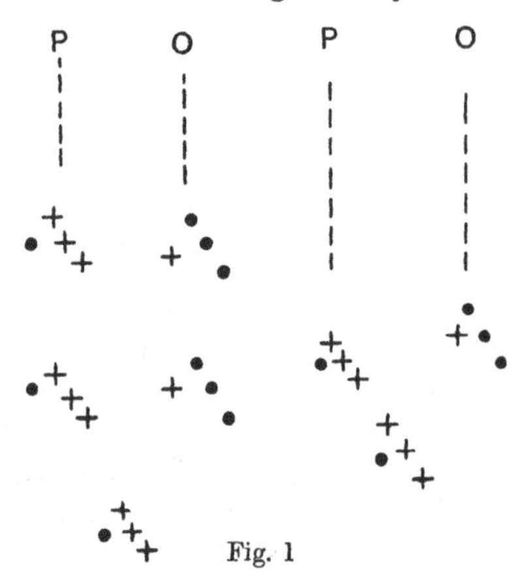

Fig. 1

Weak transitions of the order of the spin perturbation are now possible between P- and O-terms, as each contains both ●'s and +'s. Either the ● system or the + system is a complete solution of the actual problem; so that we may have either

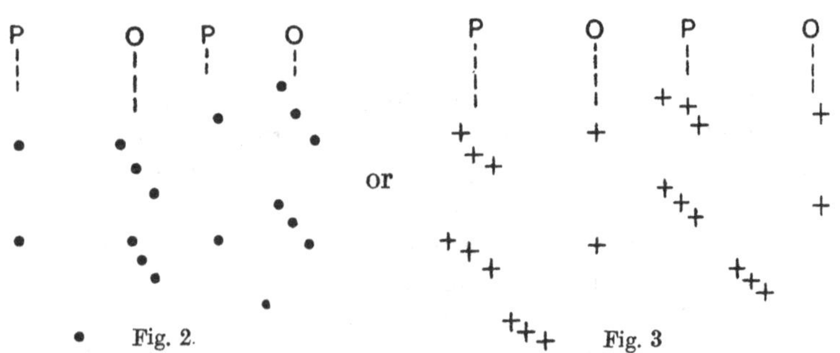

Fig. 2. Fig. 3

as representing the helium terms.

The former makes the P-terms singulets and the O-terms triplets; the latter the P-terms triplets and the O-terms singulets.

The choice between the two is settled by the Pauli verbot (§ 22), by which no two electrons can be in the same quantum state. This means that only *antisymmetric* eigenfunctions can represent a quantum state. For example the antisymmetric function used above

$$\frac{1}{\sqrt{2}}(\psi_m{}^a\psi_n{}^b - \psi_n{}^a\psi_m{}^b)$$

vanishes if the electrons a, b have the same quantum state, in which case $n = m$; so that the existence of antisymmetric eigenfunctions implies the Pauli verbot.

Finally therefore Fig. 2 represents the solution, with singulets for the P-terms and triplets for the O-terms.

136. *The calculations for helium and Li+.*

The older theories supposed one electron to be moving under the action of a nuclear charge Ze and the other electron on account of 'screening' (indicated by the K-lines) under the action of an effective nuclear charge $(Z - 1)e$ (the 'screening' constant σ_1 of § 27 is approximately 1).

But the essence of the resonance theory of Heisenberg is that the doings of each electron should be the same. He therefore supposes as his model an arrangement which near the nucleus will make either electron move under Ze and when far away move under $(Z - 1)e$. The model consists of a nucleus surrounded by a shell of radius r_0 with a charge $-e$ spread over it, so that the potential energy of either electron outside the shell is $-\frac{e^2Z}{r}+\frac{e^2}{r}$, but when it dives into the shell is $-\frac{e^2Z}{r}+\frac{e^2}{r_0}$; or,

$$-\frac{e^2Z}{r} + \text{constant},\ r < r_0, \text{and } -\frac{e^2}{r}(Z-1),\ r > r_0.$$

See also W. Heisenberg, Zs. f. Phys. **41**, p. 239, **1927**.

By the use of this model, his resonance theory and Schrödinger's eigenfunctions as normalised for hydrogen by Waller[1], he obtains numerical results[2] which have the agreement with experiment already described (§ 129).

137. *Symmetric and antisymmetric eigenfunctions* (*Dirac*[3]-*Heisenberg*[4]).

Dirac considers an atom with two electrons a, b and denotes by (mn) that state of the atom in which a is in an orbit numbered m, and b in an orbit n. The states (mn) and (nm) are physically the same, since they differ only in the interchange of the two electrons. In order to preserve the feature of the matrices that they contain only experimentally observable magnitudes, the transitions $mn \rightarrow m'n'$, $nm \rightarrow m'n'$, $mn \rightarrow n'm'$, $nm \rightarrow n'm'$ must correspond to *one* term of the matrix, since they are physically indistinguishable.

But the symmetry between a, b requires that the amplitude of the transition $mn \rightarrow m'n'$ of x_1, a coordinate of the electron a, shall be equal to that for the transition $nm \rightarrow n'm'$ of x_2, the corresponding coordinate of the electron b, so that

$$x_1(mn, m'n') = x_2(nm, n'm').$$

But if mn and nm both define the same row and column of the matrix, as supposed above, then we have the matrix equation $x_1 = x_2$.

This only means that x_1 and x_2 cannot be represented by matrices of the above type; and the same would hold for any unsymmetrical function of x_1, x_2. Hence we must conclude that only *symmetrical* functions of the coordinates of the two electrons can be represented by matrices, as, for example, the total polarisation of the atom.

[1] I. Waller, Zs. f. Phys. **38**, p. 635, **1926**.
[2] W. Heisenberg, Zs. f. Phys. **39**, p. 503, **1926**.
[3] P. A. M. Dirac, Proc. Roy. Soc. A. **112**, p. 661, **1926**.
[4] W. Heisenberg, Zs. f. Phys. **39**, p. 499, **1926**.

Now let $\psi_m{}^a$ be the eigenfunction for a in the state m and $\psi_m{}^b$ be that for b in the state n. Neglecting coupling, the ψ for the whole atom is $\psi_m{}^a\psi_n{}^b$.

So for a in the state n, and b in the state m, $\psi = \psi_n{}^a\psi_m{}^b$.

We have thus two separate eigenfunctions for the states (mn), (nm), which would if used separately give a row and column in the matrices corresponding to (mn), and another row and column corresponding to (nm). If we are to have only one row and column corresponding to (mn) and (nm), we must limit the eigenfunctions made use of to the set $\psi_{mn} \equiv \alpha_{mn}\psi_m{}^a\psi_n{}^b + \beta_{mn}\psi_m{}^b\psi_n{}^a$, where α_{mn}, β_{mn} are constants.

This set must suffice to determine the matrix representing any symmetric function f of the coordinates of the two electrons. But the component $f(mn, m'n')$ of the matrix representing f is given by

$$f(mn, m'n') = \int dq . f . \bar{\psi}_{mn}\psi_{m'n'} \qquad (\S\ 102)$$

and the f's, being symmetric, are unaffected by interchange of a and b. Hence ψ_{mn}, $\psi_{m'n'}$ must either be unaffected by interchange of a and b, or else both change sign, if the above equation is to hold.

In the former case,

$$\alpha_{mn}\psi_m{}^a\psi_n{}^b + \beta_{mn}\psi_m{}^b\psi_n{}^a = \alpha_{mn}\psi_m{}^b\psi_n{}^a + \beta_{mn}\psi_m{}^a\psi_n{}^b,$$

so that $$\alpha_{mn} = \beta_{mn}.$$

So in the latter case, $\alpha_{mn} = -\beta_{mn}$.

Hence the necessary eigenfunctions are either *symmetric* of the type $\psi_m{}^a\psi_n{}^b + \psi_m{}^b\psi_n{}^a$ or *antisymmetric* of the type $\psi_m{}^a\psi_n{}^b - \psi_m{}^b\psi_n{}^a$. Thus symmetric eigenfunctions alone, or antisymmetric ones alone, suffice to give the matrices, and so give a complete solution of the problem.

These results are evidently general. For r electrons

$(a_1, a_2, \dots a_r)$ the antisymmetric eigenfunctions may be written as a *determinant*

$$\begin{vmatrix} \psi_{n_1}^{a_1}, & \psi_{n_1}^{a_2}, & \dots & \psi_{n_1}^{a_r} \\ \psi_{n_2}^{a_1}, & \psi_{n_2}^{a_2}, & \dots & \psi_{n_2}^{a_r} \\ \dots & \dots & \dots & \dots \\ \psi_{n_r}^{a_1}, & \psi_{n_r}^{a_2}, & \dots & \psi_{n_r}^{a_r} \end{vmatrix} \qquad \dots\dots\dots\dots\dots(1)$$

and the symmetric ones written as a *permanent*, to use MacMahon's term[1], which is the determinant with all its signs positive.

If the electrons are coupled, there will still be symmetric and antisymmetric eigenfunctions, as has been seen (§ 135).

From (1), an antisymmetric eigenfunction vanishes if say $a_1 = a_2$, that is if two electrons are in the same orbit, for then two columns of the determinant become the same. This means that in the solution of the problem with antisymmetric eigenfunctions, there can be no stationary states with more than one electron in the same orbit, which is just the Pauli verbot (§ 22).

The solution with symmetric eigenfunctions, however, allows any number of electrons to be in the same orbit, and this leads to the Bose-Einstein statistics (§ 139). This must be the correct one for light quanta, for the Bose scheme leads to Planck's law of black body radiation.

[1] P. A. MacMahon, Trans. Camb. Phil. Soc. **23**, p. 122, **1924**.

CHAPTER XXVIII

THE NEW STATISTICS OF GASES AND RADIATION; THE BOSE STATISTICS FOR LIGHT QUANTA; THE EINSTEIN THEORY OF AN IDEAL GAS

138. *The new statistics of gases and radiation.*

We use the idea of de Broglie that with the motion of every atom is associated the propagation of a certain wave. The enclosure containing the gas or radiation thus contains waves of all frequencies. An atom may be either an atom of a gas (of mass m_0 with its de Broglie wave) or an atom of light or light quantum (of mass zero with only the de Broglie wave).

139. *The Bose-Einstein statistics.*

The new statistics of Bose[1] for light quanta and of Einstein[2] for a gas can thus be included in one theory, given by de Broglie[3] as follows.

Let A_s, N_s be the number of cells and atoms associated with the energy range E_s to $E_s + dE_s$; A_s, N_s are both large, but the number of atoms per cell may be any number, 0, 1, 2, 3, etc.

The number of possible partitions of the atoms into the cells is given by the formula used by Planck[4] in his earliest theory of the black body spectrum, namely

$$\frac{(A_s + N_s)\,!}{A_s\,!\,N_s\,!}\,.$$

[1] S. N. Bose, Zs. f. Phys. **27**, p. 384, **1924**.

[2] A. Einstein, Sitzungsber. der Preuss. Akad. p. 261, **1924**; p. 3, **1925**.

[3] L. de Broglie, 'Ondes et mouvements,' chap. XIII, Gauthier-Villars, Paris, **1926**.

[4] Q.T.A. § 9.

Thus the total number of microscopic states of the gas is

$$W = \Pi_s \frac{(A_s + N_s)!}{A_s!\,N_s!}.$$

The entropy

$$S = k \log W$$
$$= k\Sigma_s \{\log (A_s + N_s)! - \log A_s! - \log N_s!\}.$$

Using Stirling's theorem that $n! = \sqrt{2\pi n}\left(\frac{n}{e}\right)^n$ for large numbers, so that $\log n!$ is effectively $n \log n - n$, for large n,

$$S = k\Sigma_s \{(A_s + N_s) \log (A_s + N_s) - A_s \log A_s - N_s \log N_s\} \quad \text{.........(1)}.$$

The entropy S must be a maximum for small variations δN_s, which leave the total number of molecules N $(= \Sigma_s N_s)$ and the total energy $(= \Sigma E_s N_s)$ unaltered.

In the usual manner, we have

$$0 = \Sigma_s \{\log (A_s + N_s) - \log N_s\} \delta N_s,$$

with

$$0 = \Sigma_s \delta N_s$$
$$0 = \Sigma_s E_s \delta N_s.$$

Hence using undetermined multipliers,

$$\log (A_s + N_s) - \log N_s + \lambda + \mu E_s = 0,$$

where λ, μ are constants.

$$\therefore \quad \frac{A_s + N_s}{N_s} = e^{\alpha + \beta E_s},$$

where α, β are constants, which can be found from the conditions $N = \Sigma_s N_s$, $E = \Sigma_s E_s N_s$.

$$\therefore \quad \frac{A_s}{N_s} = e^{\alpha + \beta E_s} - 1$$

or

$$N_s = \frac{A_s}{e^{\alpha + \beta E_s} - 1} \quad \text{.....................(2)}.$$

The entropy then reduces to

$$\begin{aligned} S &= k\Sigma_s \{(A_s + N_s) \log (N_s e^{\alpha+\beta E_s}) - N_s \log N_s - A_s \log A_s\} \\ &= k\Sigma_s \{A_s (\log N_s - \log A_s) + (A_s + N_s)(\alpha + \beta E_s)\} \\ &= k\Sigma_s \{- A_s \log (e^{\alpha+\beta E_s} - 1) + A_s \log (e^{\alpha+\beta E_s}) \\ &\qquad\qquad + N_s (\alpha + \beta E_s)\} \\ &= k\Sigma_s \{N_s (\alpha + \beta E_s) - A_s \log (1 - e^{-\alpha-\beta E_s})\} \\ &= k \{N\alpha + E\beta - \Sigma_s A_s \log (1 - e^{-\alpha-\beta E_s})\}. \end{aligned}$$

To connect β with the temperature T, we have

$$\begin{aligned} \frac{1}{T} &= \frac{dS}{dE} = \frac{\partial S}{\partial E} + \frac{\partial S}{\partial \beta}\frac{d\beta}{dE} \\ &= k\beta + kE\frac{d\beta}{dE} - k\frac{d\beta}{dE}\Sigma_s \frac{A_s E_s}{e^{\alpha+\beta E_s} - 1} \\ &= k\beta + kE\frac{d\beta}{dE} - k\frac{d\beta}{dE}\Sigma N_s E_s, \text{ using (2)} \\ &= k\beta. \end{aligned}$$

$$\therefore \quad \beta = \frac{1}{kT}.$$

$$\therefore \quad S = kN\alpha + \frac{E}{T} - k\Sigma_s A_s \log (1 - e^{-\alpha-\frac{E_s}{kT}}) \quad \ldots(3)$$

and

$$N_s = \frac{A_s}{e^{\alpha+\frac{E_s}{kT}} - 1} \quad \ldots\ldots\ldots\ldots\ldots\ldots(4).$$

Equations (3), (4) are those of Einstein for an ideal gas.

If the atoms are atoms of radiation (light quanta) and the radiation is in equilibrium with a black body, the conditions are those of a gas in contact with its liquid phase, that is, the total number of atoms in the light-quantum gas is not fixed and the condition $\Sigma_s N_s = N$ must be omitted. This means that in the theory just given the constant α will not appear. We then have

$$N_s = \frac{A_s}{e^{\frac{E_s}{kT}} - 1} \quad \ldots\ldots\ldots\ldots\ldots\ldots\ldots(5),$$

which is the formula of Bose for light quanta, which suggested to Einstein his theory of an ideal gas.

140. *Bose's theory for a gas of light quanta (radiation).*

It remains to find A_s, the number of cells in the q-space associated with energy E_s.

For a light quantum, $E_s = h\nu_s$, where ν_s is the frequency; the corresponding momentum is $h\nu_s/c$, in the direction of its motion (§ 115); its state is specified by the coordinates x, y, z and momenta p_x, p_y, p_z.

The q-space is six-dimensional and the phase point lies on

$$p_x^2 + p_y^2 + p_z^2 = \frac{h^2 \nu_s^2}{c^2} \quad \text{..............(1).}$$

To a frequency range $d\nu_s$ belongs a portion of the phase space equal to

$$\iiiint\!\!\iint dx\,dy\,dz\,dp_x\,dp_y\,dp_z = V \iiint dp_x\,dp_y\,dp_z,$$

where V is the volume of the enclosure containing the radiation,

$$= V \left\{ \text{volume between two spheres of radii } \frac{h\nu_s}{c} \text{ and } \frac{h(\nu_s + d\nu_s)}{c} \right\},$$

on account of (1),

$$= V 4\pi \left(\frac{h\nu_s}{c}\right)^2 d\left(\frac{h\nu_s}{c}\right)$$

$$= 4\pi V \frac{h^3 \nu_s^2 d\nu_s}{c}.$$

Dividing up the phase space into 'cells' of volume h^3 after the manner of Planck, the number of cells corresponding to a frequency range $d\nu_s$ is

$$\frac{4\pi V \nu_s^2 d\nu_s}{c^3} \quad \text{........................(2).}$$

Light quanta however differ from gas molecules in that they are not isotropic, on account of their polarisation.

The electric vector of the quantum is always at right angles to its motion, but its orientation round it may be anything. If we suppose n atoms of light polarised in a given plane, there must be n atoms polarised in a perpendicular plane; and this is true for all planes on account of the isotropic character of the black body radiation. Thus N light-quanta are like $2N$ gas molecules. To allow for this we double the number of cells and treat the quanta like gas molecules. Therefore, allowing for the polarisation, the number of cells

$$A_s = \frac{8\pi V \nu_s^2 d\nu_s}{c^3} \quad \text{.....................(3)},$$

corresponding to the range $d\nu_s$.

$$\begin{aligned} \therefore \quad E &= \Sigma_s N_s E_s \\ &= \Sigma_s N_s h\nu_s \\ &= \Sigma_s \left(\frac{A_s}{e^{\frac{E_s}{kT}} - 1} \right) h\nu_s, \text{ using (5), § 139,} \\ &= \Sigma_s \frac{8\pi V h \nu_s^3 d\nu_s}{c^3 \left(e^{\frac{h\nu_s}{kT}} - 1\right)}, \end{aligned}$$

which is Planck's formula for black body radiation.

141. *Einstein's theory of an ideal gas.*

The number of cells A_s in the q-space associated with energy E_s is

$$\frac{V}{h^3} \iiint dp_x dp_y dp_z,$$

as before; and

$$\frac{1}{2m}(p_x^2 + p_y^2 + p_z^2) = \text{the kinetic energy} = E_s;$$

$$\therefore \quad p_x^2 + p_y^2 + p_z^2 = 2mE_s = g^2, \text{ suppose.}$$

Then $\iiint dp_x dp_y dp_z =$ volume between spheres of radii $g, g + dg$

$$= 4\pi g^2 dg.$$

$$\therefore\ A_s = \frac{V}{h^3} \cdot 4\pi g^2 dg = \frac{2\pi V}{h^3} (2m)^{\frac{3}{2}} E_s^{\frac{1}{2}} dE_s \quad \text{...(1)}.$$

Therefore, using (4), § 139,

$$N = \Sigma_s N_s = \Sigma_s \frac{A_s}{e^{\alpha + \frac{E_s}{kT}} - 1},$$

or

$$N = \frac{2\pi V}{h^3} (2m)^{\frac{3}{2}} \int_0^\infty \frac{E^{\frac{1}{2}} dE}{e^{\alpha + \frac{E}{kT}} - 1} \quad \text{............(2)}.$$

Also

$$E = \Sigma_s E_s N_s = \frac{2\pi V}{h^3} (2m)^{\frac{3}{2}} \int_0^\infty \frac{E^{\frac{3}{2}} dE}{e^{\alpha + \frac{E}{kT}} - 1} \quad \text{...(3)}.$$

The elimination of α between (2) and (3), together with the general formula $PV = \frac{2}{3} E$ of statistical mechanics, where P is the pressure, leads to the equation of state of the gas.

The entropy

$$S = kN\alpha + \frac{E}{T} - k\Sigma_s A_s \log (1 - e^{-\alpha - \frac{E_s}{kT}}), \quad (\S\ 139).$$

Using (1) this becomes

$$S = kN\alpha + \frac{E}{T} - k \frac{2\pi V}{h^3} (2m)^{\frac{3}{2}} \int_0^\infty E^{\frac{1}{2}} \log (1 - e^{-\alpha - \frac{E}{kT}})\, dE \quad \text{...(4)}.$$

142. *The classical-quantum theory as a limiting case of Einstein's theory.*

If we neglect unity compared with $e^{\alpha + \frac{E_s}{kT}}$, we obtain the results of the classical theory as adapted to the quantum theory by Planck; it will appear as we proceed for what physical conditions this approximation is exact enough.

From (2), § 141, $N = c\int_0^\infty \frac{E^{\frac{1}{2}}dE}{e^{\alpha+\frac{E}{kT}}-1}$,

where $c = \frac{2\pi V}{h^3}(2m)^{\frac{3}{2}}$, so that N

$= c\int_0^\infty E^{\frac{1}{2}} e^{-\alpha-\frac{E}{kT}}dE$, using the above approximation...(1)

$$= ce^{-\alpha}\int_0^\infty E^{\frac{1}{2}} e^{-\frac{E}{kT}}dE.$$

Since $\int_0^\infty x^{\frac{1}{2}} e^{-ax}dx = \Gamma\left(\frac{3}{2}\right) a^{-\frac{3}{2}} = \frac{1}{2}\sqrt{\pi}a^{-\frac{3}{2}}$,

$$N = ce^{-\alpha}\frac{\sqrt{\pi}}{2}(kT)^{\frac{3}{2}} \quad \text{..............(2).}$$

$$\therefore \quad e^{\alpha} = \frac{\sqrt{\pi}\,(kT)^{\frac{3}{2}}}{2N}\frac{2\pi V}{h^3}(2m)^{\frac{3}{2}},$$

inserting the value of c,

so that $$e^{\alpha} = \frac{V}{Nh^3}(2\pi mkT)^{\frac{3}{2}} \quad \text{..............(3).}$$

Also from (3), § 141,

$$E = c\int_0^\infty \frac{E^{\frac{3}{2}}dE}{e^{\alpha+\frac{E}{kT}}-1}$$

$$= c\int_0^\infty E^{\frac{3}{2}} e^{-\alpha-\frac{E}{kT}}dE,$$

again using the approximation;

$$= ce^{-\alpha}\int_0^\infty E^{\frac{3}{2}} e^{-\frac{E}{kT}}dE.$$

Since $\int^\infty x^{\frac{3}{2}} e^{-ax}dx = \Gamma\left(\frac{5}{2}\right) a^{-\frac{5}{2}} = \frac{3}{4}\sqrt{\pi}a^{-\frac{5}{2}}$,

$$E = ce^{-\alpha}\,\tfrac{3}{4}\sqrt{\pi}\,(kT)^{\frac{5}{2}} \quad \text{..................(4).}$$

The equation (4), § 141, using the same approximation, becomes

$$S = kN\alpha + \frac{E}{T} + kc\int_0^\infty E^{\frac{1}{2}} e^{-\alpha - \frac{E}{kT}} dE$$

$$= kN\alpha + \frac{E}{T} + kN, \text{ from (1).}$$

$$\therefore \quad S = \frac{E}{T} + kN(\alpha + 1).$$

But from (2) and (4), $\dfrac{E}{N} = \dfrac{3}{2}kT$,

so that
$$S = kN\left(\alpha + \frac{5}{2}\right).$$

$$\therefore \quad S = kN\left[\log\left\{\frac{V}{Nh^3}(2\pi mkT)^{\frac{3}{2}}\right\} + \frac{5}{2}\right], \text{ using (3),}$$

or
$$S = kN\log\left\{\frac{e^{\frac{5}{2}}V}{Nh^3}(2\pi mkT)^{\frac{3}{2}}\right\},$$

the formula of Planck[1].

From (3) it is found on substitution that for hydrogen at atmospheric pressure, $e^\alpha = 6 \times 10^4$ and so is large compared with unity. Hence (3), which represents the result of the classical statistics as modified for the quantum theory by Planck, gives a good approximation. But the approximation fails as the density of the gas increases or its temperature falls, and for helium near the critical state the failure is considerable.

143. Einstein has shown that his theory leads to entropy zero as $T \to 0$; and also that the specific heat of the gas increases until a maximum is reached, 'the saturation point,' and then steadily tends to zero as $T \to 0$. The details of this are not given here, as they are given for the Fermi-Dirac statistics of § 144, which probably more accurately represent the ideal gas.

[1] 'Wärmestrahlung,' by M. PLANCK, Berlin; p. 208 (5th edition), 1923.

CHAPTER XXIX

THE FERMI-DIRAC THEORY OF AN IDEAL GAS; JORDAN'S FORMULAE FOR COLLISIONS OF LIGHT QUANTA, PROTONS AND ELECTRONS

144. *The Fermi*[1]*-Dirac*[2] *statistics of an ideal gas.*

Both Fermi and Dirac base their statistics on the Pauli verbot.

Fermi generalises the Pauli rule that at most one atom with given quantum numbers can be present in a gas; the quantum numbers include not only those which fix the inner motion of the atom (those of Pauli), but also those fixing its translatory motion. He then considers a system of atoms acted upon by an elastic force to a fixed point (he assumes that the nature of the field does not affect the statistical results) so that each atom is a simple harmonic oscillator. The quantum numbers for an oscillating atom are s_1, s_2, s_3 and he omits the inner quantum numbers, supposing the atoms to be in the ground state. The Pauli rule then becomes—'there can be in the whole gas mixture at the most one atom with quantum numbers s_1, s_2, s_3.' The total energy of the atom is $h\nu\,(s_1 + s_2 + s_3)$, where after Heisenberg (§ 57) each s may have the values $\frac{1}{2}, \frac{3}{2}, \ldots$.

Therefore the total energy of the atom is $h\nu s$, where $s = s_1 + s_2 + s_3$; an atom with energy $E_s \equiv h\nu s$ is an 's-atom.'

The value $h\nu s$ can be realised in many ways; each complexion corresponds to a solution of $s_1 + s_2 + s_3 = s$, and this has $\frac{1}{2}\,(s + 1)\,(s + 2)$ solutions, $\equiv A_s$ suppose.

[1] E. FERMI, Lincei Rend. **3**, p. 145, **1926** and Zs. f. Phys. **36**, p. 902, **1926**.

[2] P. A. M. DIRAC, Proc. Roy. Soc. A. **112**, p. 661, **1926**.

Fermi then examines how energy E partitions itself among the N atoms.

If N_s is the number of s-atoms, then by the Pauli rule $N_s \leqslant A_s$, as in the gas not more than A_s s-atoms can appear.

Also
$$\left.\begin{aligned}\Sigma_s N_s &= N\\ \Sigma_s E_s N_s &= E\end{aligned}\right\} \qquad \text{.......................(1).}$$

The number of arrangements of N_s atoms in the A_s places of energy $sh\nu$ is
$$\frac{A_s!}{N_s!\,(A_s - N_s)!},$$
so that the probability
$$W = \Pi_s \frac{A_s!}{N_s!\,(A_s - N_s)!}.$$
This is to be a maximum, subject to conditions (1).

Proceeding as in § 139, we have
$$\begin{aligned}\log W &= \Sigma_s \{\log A_s! - \log N_s! - \log (A_s - N_s)!\}\\ &= \Sigma_s \{A_s \log A_s - N_s \log N_s\\ &\qquad - (A_s - N_s) \log (A_s - N_s)\} \text{...(2).}\end{aligned}$$

The maximum conditions are:
$$\left.\begin{aligned}0 &= \Sigma_s \{-\log N_s + \log (A_s - N_s)\}\, \delta N_s\\ 0 &= \Sigma_s\, \delta N_s\\ 0 &= \Sigma_s E_s\, \delta N_s\end{aligned}\right\}.$$
$$\therefore\ \log (A_s - N_s) - \log N_s + \lambda + \mu E_s = 0.$$
$$\therefore\ \frac{A_s - N_s}{N_s} = e^{\alpha + \beta E_s}.$$
$$\therefore\ N_s = \frac{A_s}{e^{\alpha + \beta E_s} + 1}.$$

β can be identified with $\frac{1}{kT}$ as in § 139 (though Fermi proceeds differently), and we have
$$N_s = \frac{A_s}{e^{\alpha + \frac{E_s}{kT}} + 1} \qquad \text{..................(3),}$$

which differs only from the Einstein formula of § 139, equation (4), in that there is a plus sign between the terms of the denominator instead of a minus sign.

We then have finally formulae corresponding to those of § 141, namely

$$\left.\begin{aligned} N &= \frac{2\pi V}{h^3}(2m)^{\frac{3}{2}}\int_0^\infty \frac{E^{\frac{1}{2}}dE}{e^{\alpha+\frac{E}{kT}}+1} \\ E &= \frac{2\pi V}{h^3}(2m)^{\frac{3}{2}}\int_0^\infty \frac{E^{\frac{3}{2}}dE}{e^{\alpha+\frac{E}{kT}}+1} \end{aligned}\right\}.$$

Fermi gives pure mathematical data which enable him to eliminate α for large or small values of T.

For *low temperatures* he finds the results:

The pressure

$$p = \frac{1}{20}\left(\frac{6}{\pi}\right)^{\frac{2}{3}}\frac{h^2 n^{\frac{5}{3}}}{m} + \frac{2^{\frac{4}{3}}\pi^{\frac{8}{3}}}{3^{\frac{5}{3}}}\cdot\frac{mn\, k^2 T^2}{h^2} + \ldots.$$

The mean kinetic energy

$$\overline{w} = \frac{3}{40}\left(\frac{6}{\pi}\right)^{\frac{2}{3}}\frac{h^2 n^{\frac{2}{3}}}{m} + \frac{2^{\frac{4}{3}}\pi^{\frac{8}{3}}}{3^{\frac{2}{3}}}\frac{mk^2T^2}{h^2 n^{\frac{2}{3}}} + \ldots,$$

where n is the number of atoms per unit volume.

Thus there is a 'nul-point' pressure and energy, as they remain finite as $T \to 0$.

Also the specific heat

$$c_v = \frac{d\overline{w}}{dT} = \frac{2^{\frac{4}{3}}\pi^{\frac{8}{3}}}{3^{\frac{2}{3}}}\frac{mk^2T}{h^2 n^{\frac{2}{3}}} + \ldots,$$

so that $c_v \to 0$ at the nul-point and for low temperatures varies as T.

For *high temperatures* he finds

$$p = nkT\left\{1 + \frac{1}{16}\frac{nh^3}{(\pi m kT)^{\frac{3}{2}}} + \ldots\right\},$$

so that the pressure is *greater* than the classical value nkT.

Einstein found

$$p = nkT\left\{1 - (\cdot 186)\frac{h^3 n^4}{(2\pi mkT)^{\frac{3}{2}}} + \dots\right\},$$

which is *less* than the classical value.

Fermi also found that the entropy at high temperatures is

$$S = nk\left[\tfrac{3}{2}\log T - \log n + \log\left\{\frac{e^{\frac{5}{2}}(2\pi mk)^{\frac{3}{2}}}{h^3}\right\}\right],$$

which agrees with the S value of Stern and Tetrode[1].

It is evident that the Fermi-Dirac theory gives the classical results of Planck in the same way as the Einstein gas theory, since the $+1$ or -1 in N_s drops out.

145. *Dirac's theory of an ideal gas.*

Dirac has independently obtained the same results as Fermi without the special assumptions as to the central field, etc., which Fermi uses; he finds them by a general method depending upon the properties of Schrödinger's eigenfunctions.

The wave equation for an atom of mass m in free space is $(H - E)\psi = 0$, where

$$H = \frac{1}{2m}(p_x{}^2 + p_y{}^2 + p_z{}^2) = \frac{1}{2m}\left[\left(-\frac{ih}{2\pi}\frac{\partial}{\partial x}\right)^2 + \dots\right]$$

and

$$E = \frac{ih}{2\pi}\frac{\partial}{\partial t}.$$

It is therefore

$$\left\{\frac{\partial^2}{\partial x^2} + \frac{\partial^2}{\partial y^2} + \frac{\partial^2}{\partial z^2} - \frac{4\pi m}{ih}\frac{\partial}{\partial t}\right\}\psi = 0.$$

Writing $\psi = ue^{-\frac{2\pi i}{h}Et}$,

$$\left\{\frac{\partial^2}{\partial x^2} + \frac{\partial^2}{\partial y^2} + \frac{\partial^2}{\partial z^2} + \frac{8\pi^2 m}{h^2}E\right\}u = 0.$$

[1] O. Stern, Phys. Zs. **14**, p. 629, **1913**.
H. Tetrode, Ann. der Phys. **38**, p. 434 and **39**, p. 255, **1912**.

The solution of this is

$$u = \exp 2\pi i\,(\alpha_1 x + \alpha_2 y + \alpha_3 z)/h,$$

where $$\alpha_1{}^2 + \alpha_2{}^2 + \alpha_3{}^2 = 2mE.$$

$$\therefore\ \left.\begin{aligned} \psi_{\alpha_1\alpha_2\alpha_3} &= \exp 2\pi i\,(\alpha_1 x + \alpha_2 y + \alpha_3 z - Et)/h \\ \text{where}\qquad \alpha_1{}^2 + \alpha_2{}^2 + \alpha_3{}^2 &= 2mE \end{aligned}\right\} \ldots(1).$$

This eigenfunction represents a 'wave' associated with the atom whose momentum components are $\alpha_1, \alpha_2, \alpha_3$ and whose energy is E.

Hence as in § 141 it follows that the number of waves associated with atoms whose energies lie between E and $E + dE$ is $\frac{2\pi V}{h^3}(2m)^{\frac{3}{2}}E^{\frac{1}{2}}dE$, where V is the volume of the enclosure. (Dirac deduces this directly from the properties of the eigenfunctions.)

He then considers the eigenfunctions of an assembly of atoms between which there is supposed to be no interaction. The possible eigenfunctions for the system are built up of products of those of the separate atoms and must be *either* a symmetric set *or* an antisymmetric set (§ 136). He makes the fundamental assumption that all the stationary states of the assembly (each state being represented by one eigenfunction) have the same *a priori* probability.

If we use symmetric eigenfunctions, then all values for the number of atoms associated with a given eigenfunction have the same probability and this leads to the Bose-Einstein statistics (§ 139).

But if we use antisymmetric eigenfunctions, then there is at most one atom associated with each eigenfunction (§ 137), which is the Pauli verbot.

The solution with symmetric eigenfunctions must be the correct one for light quanta, since it is known that the Bose-Einstein statistics lead to Planck's law of black body radiation.

The solution with antisymmetric eigenfunctions must be

the correct one for electrons, since it leads to the Pauli verbot; it is probably the correct one for gas atoms, as one would expect gas atoms to resemble electrons more closely than light quanta.

Dirac then proceeds to work out the gas theory on the assumption that the solution with antisymmetric eigenfunctions is the correct one, so that not more than one atom can be associated with each eigenfunction.

He divides the waves into a number of sets such that the waves in each set are associated with atoms of energy E to $E + dE$. Let A_s be the number of waves of the sth set and E_s the corresponding kinetic energy, so that

$$A_s = \frac{2\pi V}{h^3}(2m)^{\frac{3}{2}} E_s^{\frac{1}{2}} dE_s.$$

Then the number of antisymmetric eigenfunctions corresponding to distributions in which N_s atoms are associated with waves of the sth set is

$$\Pi_s \frac{A_s!}{N_s!\,(A_s - N_s)!};$$

and this is the probability W, giving the entropy S $(= k \log W)$.

The rest of the procedure is the same as that of Fermi (§ 144) and leads to the same results.

The saturation phenomenon of the Einstein theory does not occur on this theory. The specific heat tends steadily to zero as $T \to 0$ and does not rise to a maximum and then fall in the manner described in § 144.

A critical discussion of the present state of quantum theory statistics has lately been given by Fowler[1].

146. *Jordan's analysis of the new statistics.*

Jordan[2] has lately given an illuminating analysis of the Bose-Einstein and Fermi-Dirac statistics in relation to the thermodynamical equilibrium of cosmic matter.

[1] R. H. Fowler, Proc. Roy. Soc. A. **113**, p. 432, 1927.
[2] P. Jordan, Zs. f. Phys. **41**, p. 711, 1927.

A formula for the thermal equilibrium concentration of matter had been found by Stern[1], who used the idea of J. J. Thomson[2] that electrons and protons can neutralise one another and become radiation. Jordan proceeds to show how Stern's formula can be deduced from the Bose-Einstein or the Fermi-Dirac theory.

147. *On the Bose-Einstein theory,* if n is the mean number of atoms (or light quanta) per cell in the phase space, then the entropy per cell is

$$S = k \log W = k\{(1+n)\log(1+n) - n \log n\},$$

as is seen from equation (1), § 139.

Jordan considers the following cases:

(A) the 'rest' energy of a particle is zero (light quanta);

(B) the 'rest' energy of a particle is positive (gas atoms);

and further

(a) the total number of particles is variable;

(b) the total number of particles is constant.

In the cases (a) we find the maximum of the integral of S for all the cells subject to the total energy being constant; in the cases (b) this maximum is found with the additional condition that the integral of n for all the cells is constant.

In cases (a) we have, as in § 139, equation (5),

$$n = \frac{1}{e^{\frac{E}{kT}} - 1} \quad \text{.....................(1)},$$

and in cases (b) we have, as in § 139, equation (4),

$$n = \frac{1}{e^{a+\frac{E}{kT}} - 1} \quad \text{.....................(2)},$$

[1] O. STERN, Zs. f. Elektrochem. **31**, p. 448, **1925**; Zs. f. phys. Chem. **120**, p. 60, **1926**.

[2] J. J. THOMSON, Phil. Mag. **48**, p. 737, **1924**.

where α is a function of T found from the condition that the integral of n is constant.

Case (A, a) is the Bose formula for light quanta in equilibrium with matter and leads as has been seen (§ 140) to Planck's radiation formula.

Case (A, b) is a formula for light quanta in a reflecting enclosure in which there is no matter which can absorb or emit radiation; variations of frequency of the light quanta may occur through the Compton effect, but no variation in their total number.

This formula would lead to a modified form of Planck's formula for radiation under these conditions.

Case (B, a) is the case of a gas in contact with its liquid phase.

Case (B, b) is that of the Einstein gas theory.

148. *On the Fermi-Dirac theory*, the formula for S is

$$S = k\{n \log n - (1 - n) \log (1 - n)\},$$

as is seen from equation (2), § 144; and in place of (1), (2) of § 147, we find

$$n = \frac{1}{e^{\frac{E}{kT}} + 1} \quad \text{.....................(3)},$$

and

$$n = \frac{1}{e^{\alpha + \frac{E}{kT}} + 1} \quad \text{.....................(4)},$$

where α is a different function of T from the α in (2), § 147 (cf. equation (3), § 144).

These results follow from the Pauli verbot and would apply to an electron gas, the former when the number of electrons is unlimited and the latter when the number is constant. The latter is the form used by Fermi and Dirac for an ideal gas when the number of gas molecules is constant.

149. *Deduction of Stern's formula.*

This follows from either of the formulae $n = \dfrac{1}{e^{\frac{E}{kT}} \pm 1}$ of the two schemes. Writing $E = mc^2 + \frac{1}{2}mv^2$, we have, since the first factor of the denominator $\gg 1$,

$$n = e^{-\frac{E}{kT}} = e^{-\frac{mc^2}{kT}-\frac{mv^2}{2kT}}, \text{ the Stern result.}$$

150. *Jordan's application of the statistics to a system of electrons and protons.*

Since the total charge of the system is constant, the difference of the number of protons and the number of electrons is constant.

Jordan constructs for the protons and for the electrons two cell partitions in the phase space.

Using first the Bose-Einstein theory, we have in the above notation

$$S' = k\{(1+n')\log(1+n') - n'\log n'\},$$
$$S'' = k\{(1+n'')\log(1+n'') - n''\log n''\},$$

where S', n' refer to the electrons and S'', n'' to the protons. The total entropy, which is the sum of the integrals of S' and S'' taken for all their cells, is to be a maximum, subject to constancy of the integrals of $n' - n''$ and of $n'E' - n''E''$.

This leads to

$$\left.\begin{aligned} n' &= \frac{1}{e^{\alpha+\frac{E'}{kT}} - 1} \\ n'' &= \frac{1}{e^{-\alpha+\frac{E''}{kT}} - 1} \end{aligned}\right\},$$

where α is the same function of T in the two expressions.

Had we used the Fermi-Dirac theory the same formulae would have resulted except that $+\ 1$ would have taken the place of $-\ 1$ in the denominator.

Using a suggestion by Hund that for an electron the Pauli verbot, and with it the Fermi-Dirac statistics, holds, but that for a proton the Bose-Einstein statistics should be used, Jordan writes

$$\left.\begin{aligned} n^{-} &= \frac{1}{e^{\alpha+\frac{E}{kT}}+1}, \text{ for an electron} \\ n^{+} &= \frac{1}{e^{-\alpha+\frac{E}{kT}}-1}, \text{ for a proton} \end{aligned}\right\} \qquad \text{.........(1).}$$

151. *Jordan's theory of the equilibrium of two gases.*

Each gas has its cell partition in the phase space. Consider any pair of cells of the first gas, which can be numbered 1 and 3. Let n_1, n_3 be the number of atoms and E_1, E_3 the energies associated with these two cells.

Consider collisions which cause a phase point of this gas to move from cell 1 to cell 3. Simultaneously a phase point of the second gas will move in its phase space from a cell which we will number 2 to a cell we will number 4. If such a collision is described as 'direct,' then one which causes a phase point of the first gas to move from cell 3 to cell 1 and a phase point of the second gas from cell 4 to cell 2, is described as 'inverse.'

The total energy is conserved, so that

$$E_1 + E_2 = E_3 + E_4 \quad \text{..................(1).}$$

The number of collisions per second between the n_1 atoms of cell 1 and the n_2 atoms of cell 2 is proportional to $n_1 n_2$ and of these a certain proportion $a n_1 n_2$ are of the 'direct' kind. So the number of collisions per second of the 'inverse' kind is $a' n_3 n_4$.

For equilibrium

$$a n_1 n_2 = a' n_3 n_4 \quad \text{.....................(2).}$$

The old statistics gave

$$n_1 = Ce^{-\frac{E_1}{kT}},\ n_3 = Ce^{-\frac{E_3}{kT}}$$

and

$$n_2 = C'e^{-\frac{E_2}{kT}},\ n_4 = C'e^{-\frac{E_4}{kT}},$$

so that (2) became

$$ae^{-\frac{(E_1+E_2)}{kT}} = a'e^{-\frac{(E_3+E_4)}{kT}},$$

or $a = a'$, on account of (1).

This is the classical result of Boltzmann.

If however we use the Fermi-Dirac statistics, then

$$n_1 = \frac{1}{e^{\alpha+\frac{E_1}{kT}}+1},\quad n_3 = \frac{1}{e^{\alpha+\frac{E_3}{kT}}+1},$$

$$n_2 = \frac{1}{e^{\alpha'+\frac{E_2}{kT}}+1},\quad n_4 = \frac{1}{e^{\alpha'+\frac{E_4}{kT}}+1}.$$

Therefore $\left(\frac{1}{n_1}-1\right)\left(\frac{1}{n_2}-1\right) = e^{\alpha+\alpha'+\frac{E_1+E_2}{kT}}$

and $\left(\frac{1}{n_3}-1\right)\left(\frac{1}{n_4}-1\right) = e^{\alpha+\alpha'+\frac{E_3+E_4}{kT}}.$

Since $E_1 + E_2 = E_3 + E_4$, it follows that

$$\left(\frac{1}{n_1}-1\right)\left(\frac{1}{n_2}-1\right) = \left(\frac{1}{n_3}-1\right)\left(\frac{1}{n_4}-1\right).$$

Using this and (2), we have

$$\frac{a'}{a} = \frac{n_1 n_2}{n_3 n_4} = \frac{(1-n_1)(1-n_2)}{(1-n_3)(1-n_4)}.$$

Therefore we write

$$a = A(1-n_3)(1-n_4)$$

and

$$a' = A(1-n_1)(1-n_2),$$

where in the former A is independent of n_1, n_2 and in the latter of n_3, n_4, so that A is independent of the n's.

Thus the number of direct collisions per second is

$$an_1 n_2 = An_1 n_2 (1-n_3)(1-n_4),$$

and of the inverse

$$An_3n_4(1-n_1)(1-n_2).$$

(n_1, n_2 refer to atoms before, n_3, n_4 to atoms after 'direct' collision.) This is readily generalised for more than two gases.

152. *Extension to the case of a gas and radiation.*

This is the case of two gases, where one is an ideal gas and the other a light quantum gas.

For the ideal gas (supposed to be the first gas of the preceding article) we use the Fermi-Dirac statistics and write

$$n_1 = \frac{1}{e^{\alpha+\frac{E_1}{kT}}+1}, \quad n_3 = \frac{1}{e^{\alpha+\frac{E_3}{kT}}+1}.$$

For the radiation (light quantum gas) we use the Bose statistics and write

$$n_2 = \frac{1}{e^{\frac{E_2}{kT}}-1}, \quad n_4 = \frac{1}{e^{\frac{E_4}{kT}}-1}.$$

From these it follows in the same way as before that

$$\left(\frac{1}{n_1}-1\right)\left(\frac{1}{n_2}+1\right)=\left(\frac{1}{n_3}-1\right)\left(\frac{1}{n_4}+1\right)$$

and the number of 'direct' collisions per second is

$$An_1n_2(1-n_3)(1+n_4)$$

and inversely

$$An_3n_4(1-n_1)(1+n_2).$$

For a material gas $e^{\alpha+\frac{E}{kT}}$ is large, so that n_1 and n_3 are small compared with unity.

The first of these results then becomes

$$An_1n_2(1+n_4).$$

But

$$n_2 = \frac{1}{e^{\frac{E_2}{kT}}-1} = \frac{1}{e^{\frac{h\nu}{kT}}-1},$$

where ν is the frequency of the quantum before the collision, and

$$n_4 = \frac{1}{e^{\frac{E_4}{kT}} - 1} = \frac{1}{e^{\frac{h\nu'}{kT}} - 1},$$

where ν' is the frequency after the collision.

Therefore the probability of the 'direct' collision is

$$An_1 n_2 (1 + n_4)$$

$$= An_1 \frac{e^{\frac{h\nu'}{kT}}}{(e^{\frac{h\nu}{kT}} - 1)(e^{\frac{h\nu'}{kT}} - 1)},$$

which agrees with the formula of Pauli[1].

Thus the probability of a Compton collision (§ 118) for an electron gas and radiation depends upon the nature of the quanta resulting from the collision.

153. *The collision of protons and electrons.*

Jordan considers the problem of Jauncey and Hughes[2] in which a proton and two electrons by collision give rise to an electron and a light quantum[3].

Using n^+, n_1^-, n_2^- for the proton and two electrons, and n^-, σ for the resulting electron and light quantum, then we have

$$n^+ = \frac{1}{e^{-\alpha + \frac{E^+}{kT}} - 1}, \quad n^- = \frac{1}{e^{\alpha + \frac{E^-}{kT}} + 1}, \quad \text{using (1) § 150,}$$

$$n_1^- = \frac{1}{e^{\alpha + \frac{E_1^-}{kT}} + 1}, \quad n_2^- = \frac{1}{e^{\alpha + \frac{E_2^-}{kT}} + 1}.$$

$$\sigma = \frac{1}{e^{\frac{h\nu}{kT}} - 1}.$$

[1] W. Pauli, Zs. f. Phys. **18**, p. 272, **1923**; **22**, p. 261, **1924**.

[2] G. E. M. Jauncey and A. L. Hughes, Proc. Nat. Acad. Amer. **12**, p. 169, **1926**.

[3] J. J. Thomson, Phil. Mag. **48**, p. 737, **1924**.

Proceeding as in §§ 151–2,

$$an^+n_1^-n_2^- = a'n^-\sigma.$$

$$\therefore \left(\frac{1}{n^+}+1\right)\left(\frac{1}{n_1^-}-1\right)\left(\frac{1}{n_2^-}-1\right) = \exp\left\{\alpha + \frac{(E^+ + E_1^- + E_2^-)}{kT}\right\}$$

and $$\left(\frac{1}{n^-}-1\right)\left(\frac{1}{\sigma}+1\right) = \exp\left\{\alpha + \frac{E^- + h\nu}{kT}\right\}.$$

But since $E^+ + E_1^- + E_2^- = E^- + h\nu$, the energy being conserved,

$$\left(\frac{1}{n^+}+1\right)\left(\frac{1}{n_1^-}-1\right)\left(\frac{1}{n_2^-}-1\right) = \left(\frac{1}{n^-}-1\right)\left(\frac{1}{\sigma}+1\right).$$

Therefore, just as in § 151,

$$a = A(1-n^-)(1+\sigma),$$
$$a' = A(1+n^+)(1-n_1^-)(1-n_2^-),$$

and the probability of the direct collision is

$$An^+n_1^-n_2^-(1-n^-)(1+\sigma)$$

and of the inverse

$$An^-\sigma(1+n^+)(1-n_1^-)(1-n_2^-).$$

Jordan has generalised this procedure so as to include problems concerning chemical constants, thermal ionisation and the like.

CHAPTER XXX

DIRAC'S THEORY

154. *Dirac's*[1] *theory of perturbations.*

Dirac considers an atomic system disturbed by an external electromagnetic field.

The wave equation for the undisturbed system is

$$(H - E)\,\psi = 0 \quad \text{.................(1)},$$

where H is the operator $H\left(-\frac{ih}{2\pi}\frac{\partial}{\partial q}, q\right)$ and E is the operator $\frac{ih}{2\pi}\frac{\partial}{\partial t}$ (§ 105). Its general solution is $\psi = \Sigma c_n \psi_n$, where the c_n's are arbitrary constants.

He supposes the ψ_n's to be chosen so that one is associated with each stationary state (of energy E_n) of the atom; and that the c_n's are chosen so that ψ represents an assembly of the undisturbed atoms in which $|c_n|^2$ is the number of atoms in the nth state.

Now suppose a perturbation applied beginning at time $t = 0$. The wave equation for the perturbed system will be of the form

$$(H - E + V)\,\psi = 0 \quad \text{...............(2)},$$

where V is the perturbing energy, and is a function of p, q, t.

It will now be shown that a solution of (2) can be found of the form $\psi = \Sigma a_n \psi_n$, where the a_n's are functions of t only, and this ψ represents an assembly of the disturbed atoms in which $|a_n(t)|^2$ is the number in the nth state at time t.

[1] P. A. M. Dirac, Proc. Roy. Soc. A. **112, p. 661, 1926.**

Substituting in (2) we have

$$\Sigma\,(H - E + V)\,a_n\psi_n = 0.$$

Since a_n is a function of t only, H and V commute with a_n;

$$\therefore\quad \Sigma\,(a_n H - E a_n + a_n V)\,\psi_n = 0.$$

Also $$E a_n - a_n E = \frac{ih}{2\pi}\,\dot{a}_n \text{ (§ 55).}$$

$$\therefore\quad \Sigma a_n\,(H - E + V)\,\psi_n = \frac{ih}{2\pi}\,\Sigma \dot{a}_n \psi_n.$$

Therefore, since $(H - E)\,\psi_n = 0$,

$$\Sigma_n a_n V \psi_n = \frac{ih}{2\pi}\,\Sigma_n \dot{a}_n \psi_n.$$

Suppose $V\psi_n$ expanded in the form

$$V\psi_n = \Sigma_m V_{mn}\psi_m,$$

where the V_{mn}'s are functions of t only and are the matrix constituents of V (§ 101).

$$\therefore\quad \frac{ih}{2\pi}\,\Sigma_m \dot{a}_m \psi_m = \Sigma_n \Sigma_m a_n V_{mn} \psi_m.$$

Taking out the coefficient of ψ_m we have

$$\frac{ih}{2\pi}\,\dot{a}_m = \Sigma_n a_n V_{mn} \quad \text{..................(3).}$$

This is a fundamental equation which determines the a's in terms of t.

As a first approximation a_n may be put equal to the constant c_n, the value of a_n at time $t = 0$. This first approximation may be inserted for a_n on the right and a second approximation found by another integration.

Further, if N_m is the number of atoms in the nth state,

$$N_m = |\,a_m\,|^2 = a_m a_m{}^*.$$

$$\therefore\quad \frac{ih}{2\pi}\,N_m = \frac{ih}{2\pi}\,(\dot{a}_m a_m{}^* + a_m \dot{a}_m{}^*).$$

But $$\frac{ih}{2\pi}\dot{a}_m = \Sigma_n a_n V_{mn}.$$

$$\therefore\ -\frac{ih}{2\pi}\dot{a}_m{}^* = \Sigma_n a_n{}^* V_{mn}{}^*$$
$$= \Sigma_n a_n{}^* V_{nm},$$

since V is a Hermite matrix.

$$\therefore\ \frac{ih}{2\pi}\dot{N}_m = \Sigma_n (a_n V_{mn} a_m{}^* - a_n{}^* V_{nm} a_m).$$

$$\therefore\ \frac{ih}{2\pi}\Sigma_m \dot{N}_m = \Sigma_{nm} (a_m{}^* V_{mn} a_n - a_n{}^* V_{nm} a_m)$$
$$\equiv 0.$$

[a_n, V_{mn}, ... commute, since all are functions of t.]

Therefore the total number of atoms is constant, as it should be. Thus the assumption $N_m = | a_m |^2$ satisfies all the necessary conditions.

155. *Calculation of the Einstein B coefficients.*

Dirac then considers the incidence of plane polarised electromagnetic radiation upon the atomic system described above.

The perturbing term V is known in terms of the total polarisation and the potentials of the incident radiation. The a's are then found by the approximation method just described, and thence N_m, the number of atoms in the state m at time T.

In order to obtain results independent of the initial phases of the atoms, $c_n e^{i\gamma_n}$ is written for each c_n in the result for N_m and the average over all values of γ_n from 0 to 2π is taken. As a result the first order terms in N_m vanish and the second order terms give a result of the form $\Sigma_n f(n, m)$. Hence ΔN_m, the increase in the number of atoms in the state m from time $t = 0$ to $t = T$ is $\Sigma_n f(n, m)$. Thus the term $f(n, m)$ may be regarded as the part of ΔN_m due to transitions from the state n and the state m.

By averaging over all directions and states of polarisation of the incident radiation, the number of transitions from the state n to the state m is found to be

$$\frac{2\pi}{3h^2c}\{|c_n|^2 - |c_m|^2\} \, |P_{nm}|^2 . I_\nu,$$

where I_ν is the intensity of frequency ν per unit frequency range of the incident radiation (found by resolving it into its Fourier components) and P_{nm} is the matrix component of the total polarisation (with time factor omitted) corresponding to the transition $n \to m$.

Thus the radiation has caused

$$\frac{2\pi}{3h^2c} |c_n|^2 \, |P_{nm}|^2 I_\nu$$

transitions from state n to state m, and

$$\frac{2\pi}{3h^2c} |c_m|^2 \, |P_{nm}|^2 I_\nu$$

transitions from state m to state n, the probability coefficient for either process being

$$B_{n\to m} = B_{m\to n} = \frac{2\pi}{3h^2c} |P_{nm}|^2,$$

in agreement with Einstein's theory[1].

This method fails to give the Einstein A coefficients for spontaneous emission.

156. *Dirac's theory of the emission and absorption of radiation*[2].

He considers an atom interacting with radiation supposed in an enclosure so as to have only a discrete set of degrees of freedom. Resolving the radiation into its Fourier components, we can consider the energy and

[1] Q.T.A. § 26.
[2] P. A. M. Dirac, Proc. Roy. Soc. A. **114**, p. 243, **1927**.

phase of each as dynamical variables describing the radiation field.

If E_s, θ_s are the energy and phase of an s component of the Fourier series, the phase θ_s meaning the time since the wave was in a standard phase, E_s and θ_s can be taken as a pair of canonical variables. For if there is no interaction between the field and atom (the unperturbed state), the Hamiltonian $H = \Sigma E_s + H_0$, where H_0 refers to the atom alone, since the variables E_s, θ_s obviously satisfy the canonical equations of motion

$$\dot{E}_s = -\frac{\partial H}{\partial \theta_s} = 0, \quad \dot{\theta}_s = \frac{\partial H}{\partial E_s} = 1.$$

They also satisfy the quantum condition

$$-\theta_s E_s + E_s \theta_s = \frac{ih}{2\pi} 1 \quad \text{..................(1).}$$

The interaction is accounted for by the addition to H of a perturbation term, which is a function of the atom variables and the E_s, θ_s. This gives the effect of the radiation on the atom and the reaction of the atom on the field.

The quantum condition (1) immediately gives light quantum properties to the radiation. For if ν_s is the frequency of a component s, $2\pi\nu_s\theta_s$ is an angle variable, so that its canonical conjugate E_s/ν_s (remembering the Bohr relation $dE = \Sigma\, \nu dI$)[1] can only assume values differing by nh. Hence δE_s can only be $nh\nu_s$.

Thus by making the energies and phases q-numbers, Dirac obtained not only the Einstein B coefficients, but also the A coefficients.

In the earlier theory of § 155 where the energies and phases were c-numbers, only the B's could be obtained and the reaction of the atom on the radiation could not be taken into account.

[1] Q.T.A. §§ 31, 53.

Dirac first treats the problem of an assembly of similar systems satisfying the Einstein-Bose statistical mechanics, which interact with another different system, a Hamiltonian being found to represent the motion. The theory is then applied to the interaction of an assembly of light quanta with an atom and Einstein's laws for the emission and absorption of radiation are deduced. The interaction of the atom with electromagnetic waves is then considered on the lines indicated above and the Hamiltonian takes the same form as in the light quantum treatment. Thus there is complete harmony between the wave and light quantum account of the interaction.

A corresponding theory for the interaction with an atom of an assembly satisfying the Fermi-Dirac statistics has recently been given by Jordan[1].

157. *Dirac's theory of dispersion*[2].

In the original theory of Kramers and Heisenberg[3] it is assumed that the matrix elements of the polarisation of an atom determine the emission and absorption of radiation analogously to the Fourier components in the classical theory.

In later theories[4,5] the Schrödinger electric density (§ 103) is used to determine the emitted radiation by the same formulae as in the classical theory.

Dirac applies to the dispersion problem the method of the previous article in which a field of radiation can be treated as a dynamical system whose interaction with an ordinary atomic system may be described by means of a Hamiltonian.

[1] P. Jordan, Zs. f. Phys. **44**, p. 473, **1927**.
[2] P. A. M. Dirac, Proc. Roy. Soc. A. **114**, p. 710, **1927**.
[3] H. A. Kramers and W. Heisenberg, Zs. f. Phys. **31**, p. 681, **1925**. Q.T.A. §§ 148–154.
[4] E. Schrödinger, Ann. der Phys. **81**, p. 109, **1926**, or chap. XXII of this book.
[5] O. Klein, Zs. f. Phys. **41**, p. 407, **1927**.

He obtains the Kramers-Heisenberg formula and finds certain results for the case of resonance (where the incident frequency coincides with that of an absorption line), namely, that nearly all the scattered radiation is due to absorptions and emissions governed by Einstein's laws.

It suffices for this purpose to treat the atom as a dipole, but it would seem that in the application of the theory to the problem of the *breadth* of a spectral line convergence difficulties arise which would disappear if a higher order of approximation than the dipole were used.

CHAPTER XXXI

THE DIRAC MATRIX TRANSFORMATION THEORY; DEDUCTION OF SCHRÖDINGER'S WAVE EQUATION; GENERALISED FORM OF THE WAVE EQUATION; PHYSICAL INTERPRETATION OF q-NUMBERS; INTRODUCTION OF PROBABILITIES; APPLICATION TO COLLISION PROBLEMS

158. *Dirac's*[1] *generalised matrix theory.*

The principles of the theory can be completely understood by considering a dynamical system with one degree of freedom; this reduces the detail without affecting the essentials of the argument, and the extension to more degrees of freedom is fairly obvious.

In matrix mechanics a dynamical variable is denoted by a matrix g whose rows and columns refer to stationary states of the system; these are numbered by specified values $\alpha, \alpha', \alpha'', \ldots$ of some constant of integration x of the dynamical system, so that $g(\alpha\alpha')$ is a typical constituent of the matrix g. If for example, the constant of integration x is an action variable, the values $\alpha, \alpha', \alpha'', \ldots$ form a discrete set; if x is a momentum or an energy, they may have a continuous range of values, a discrete set, or both. x is a q-number whose 'characteristic values' are $\alpha, \alpha', \alpha'', \ldots$ or in other words, x is a diagonal matrix whose diagonal terms are $\alpha, \alpha', \alpha'', \ldots$.

The case where the α's form a continuous range is more general, so that this form will be used.

The matrix law of multiplication

$$ab(\alpha\alpha') = \Sigma_{\alpha''}\, a(\alpha\alpha'')\, b(\alpha''\alpha')$$

[1] P. A. M. DIRAC, Proc. Roy. Soc. A. **113**, p. 621, **1927**.

for discrete sets of α's now reads

$$ab\,(\alpha\alpha') = \int a\,(\alpha\alpha'')\,d\alpha''\,b\,(\alpha''\alpha'),$$

where the integration is taken over all the values of α'' in the continuous range.

159. *Properties of* $\delta\,(x)$.

In the theory of matrices with continuous ranges of rows and columns it is necessary to introduce the function $\delta\,(x)$, which is zero except when x is very small and whose integral through a range including $x = 0$ is unity. (It corresponds to $\delta\,(mn)$ of discrete set theory.)

We have $\delta\,(x) = 0,\ x \neq 0$ and $\int_{-\infty}^{\infty} \delta\,(x)\,dx = 1$, corresponding to

$$\begin{aligned}\delta\,(mn) &= 0,\ n \neq m\\ &= 1,\ n = m.\end{aligned}$$

We can take $\delta\,(x)$ to be symmetrical about zero, so that

$$\delta\,(-x) = \delta\,(x) \quad \text{and} \quad \delta'\,(-x) = -\,\delta\,(x), \ldots$$

where dashes denote differential coefficients. The condition $\delta\,(x) = 0,\ x \neq 0$ can be expressed by the equation $x\delta\,(x) = 0$.

From the above it follows at once that

$$\int_{-\infty}^{\infty} f\,(x)\,\delta\,(a - x)\,dx = f\,(a) \quad \ldots\ldots\ldots\ldots(1).$$

Again,

$$\int_{-\infty}^{\infty} f\,(x)\,\delta'\,(a - x)\,dx$$

$$= \left[-f\,(x)\,\delta\,(a - x)\right]_{-\infty}^{\infty} + \int_{-\infty}^{\infty} f'\,(x)\,\delta\,(a - x)\,dx$$

$= f'\,(a)$, since the first term vanishes at both limits.

$$\therefore \int_{-\infty}^{\infty} f\,(x)\,\delta'\,(a - x)\,dx = f'\,(a) \quad \ldots\ldots\ldots(2).$$

Writing $\delta(x-b)$ for $f(x)$ in (1) and (2), we have

$$\int_{-\infty}^{\infty} \delta(a-x)\,\delta(x-b)\,dx = \delta(a-b) \quad \ldots\ldots(3)$$

and
$$\int_{-\infty}^{\infty} \delta'(a-x)\,\delta(x-b)\,dx = \delta'(a-b) \ldots\ldots(4).$$

Further the unit matrix must have the property that when multiplied by any matrix y the product is equal to y, or

$$\int 1\,(\alpha\alpha')\,d\alpha' y\,(\alpha'\alpha'') = y\,(\alpha\alpha'').$$

On account of (1), this means that $1\,(\alpha\alpha') = \delta(\alpha-\alpha')$.

The general diagonal matrix f has elements

$$f(\alpha\alpha)\,\delta(\alpha-\alpha') \text{ or } f(\alpha)\,\delta(\alpha-\alpha'),$$

so that $f(\alpha), f(\alpha'), \ldots$ form the diagonal terms and zeros the rest.

160. *Matrix scheme for a dynamical problem not unique.*

The solution of a problem in Heisenberg's matrix mechanics consists in finding a scheme of matrices to represent the dynamical variables, which (for one degree of freedom) satisfies the conditions:

(i) The quantum condition $qp - pq = \frac{ih}{2\pi}\,1$.

(ii) The equations of motion $gH - Hg = \frac{ih}{2\pi}\,\dot{g}$, where g is a typical variable.

(iii) The Hamiltonian H is a diagonal matrix.

(iv) The matrices are of Hermite type.

There are any number of matrix schemes which satisfy these conditions. For if to each matrix g, say, we apply the canonical transformation $G = SgS^{-1}$, where $S\tilde{S}^* = 1$ and S is independent of the time, the new matrix G satisfies the conditions (i) to (iv) (cf. §§ 58, 59). Thus the new matrices G are just as effective as the matrices g to solve the problem.

Dirac has shown how to select from this multiplicity of matrix schemes the one which makes any desired dynamical variable ξ a *diagonal matrix*. If the constituents (diagonal) of ξ are $\beta, \beta', \beta'', \ldots$, the c-numbers $\beta, \beta', \ldots$ are the 'characteristic values' of the q-number ξ, and so by this scheme questions which have to do with *numerical values* associated with the variable denoted by the q-number can be answered, as the β's denote its possible numerical values.

161. *The transformation equations.*

Let the matrix $g \equiv (g\,(\alpha\alpha'))$ be transformed to G by the canonical transformation

$$G = SgS^{-1} \quad\ldots\ldots\ldots\ldots\ldots\ldots\ldots(1),$$

where $$S\tilde{S}^* = 1.$$

Let the rows and columns of G be numbered by

$$\beta, \beta', \beta'', \ldots,$$

so that $G\,(\beta\beta')$ is a typical constituent of the new matrix.

Dirac uses the same symbol g for both G and g, as which is meant is clearly indicated by whether it is followed by α's or β's.

From (1) we have

$$g\,(\beta\beta') = \iint S\,(\beta\alpha)\,d\alpha\, g\,(\alpha\alpha')\,d\alpha' S^{-1}\,(\alpha'\beta').$$

The transformation functions $S\,(\beta\alpha)$, $S^{-1}\,(\alpha\beta)$ will be written just as (β/α), (α/β) for brevity, so that we have

$$g\,(\beta\beta') = \iint (\beta/\alpha)\,d\alpha\, g\,(\alpha\alpha')\,d\alpha'\,(\alpha'/\beta') \quad\ldots\ldots(2).$$

But (1) may be written in the forms

$$S^{-1}GS = g, \quad GS = Sg, \quad S^{-1}G = gS^{-1}$$

which give

$$g\,(\alpha\alpha') = \iint (\alpha/\beta)\,d\beta\, g\,(\beta\beta')\,d\beta'\,(\beta'/\alpha') \quad\ldots\ldots(3),$$

$$\int g\,(\beta\beta')\,d\beta'\,(\beta'/\alpha) = \int (\beta/\alpha')\,d\alpha'\, g\,(\alpha'\alpha) \equiv g\,(\beta\alpha), \text{ say} \ldots(4),$$

$$\int (\alpha/\beta')\,d\beta'\, g(\beta'\beta) = \int g\,(\alpha\alpha')\,d\alpha'\,(\alpha'/\beta) \equiv g\,(\alpha\beta), \text{ say}\ldots(5).$$

We thus obtain two new quantities $g(\beta\alpha)$ and $g(\alpha\beta)$ which give two matrix schemes for g, where the rows and columns refer to different things; there is now no one-to-one correspondence between the rows and columns, so that a diagonal matrix in these new schemes has no meaning. The matrices whose elements are (β/α) and (α/β) are the unit matrices in the respective schemes, for equations (4), (5) show that these matrices when multiplied by a matrix representing an arbitrary q-number g give matrices representing g.

The matrices S and S^{-1} satisfy the relations $SS^{-1} = 1$ and $S^{-1}S = 1$, so that

$$\int (\beta/\alpha)\, d\alpha\, (\alpha/\beta') = \delta(\beta - \beta')$$

and $$\int (\alpha/\beta)\, d\beta\, (\beta/\alpha') = \delta(\alpha - \alpha').$$

Thus the matrix elements (β/α) and (α/β) form two orthogonal and normalised systems of functions, whether they are regarded as functions of the α's specified by values of the parameter β or as functions of the β's specified by values of the parameter α.

162. *Canonical variables ξ, η, where ξ is a diagonal matrix.*

Let ξ be the diagonal matrix

$$\begin{pmatrix} \beta & 0 & 0 & \cdots \\ 0 & \beta' & 0 & \cdots \\ 0 & 0 & \beta'' & \cdots \end{pmatrix},$$

where the β's are those introduced in § 161, so that

$$\xi(\beta\beta') = \beta\delta(\beta - \beta').$$

It will now be shown that the matrix η, where

$$\eta(\beta\beta') = -\frac{ih}{2\pi}\delta'(\beta - \beta'),$$

is canonically conjugate to ξ.

It is necessary to show that

$$\xi\eta - \eta\xi = \frac{ih}{2\pi} 1.$$

The $(\beta\beta')$ constituent of the left-hand side, namely

$$(\xi\eta - \eta\xi)(\beta\beta') = \int \{\xi(\beta\beta'')\, d\beta''\eta(\beta''\beta') - \eta(\beta\beta'')\, d\beta''\xi(\beta''\beta')\}$$

$$= -\frac{ih}{2\pi}\int \{\beta\delta(\beta-\beta'')\delta'(\beta''-\beta') - \delta'(\beta-\beta'')\beta''\delta(\beta''-\beta')\} d\beta''.$$

Integrating the second term by parts, this expression

$$= -\frac{ih}{2\pi}\int \{\beta\delta(\beta-\beta'')\delta'(\beta''-\beta') - \delta(\beta-\beta'')\frac{\partial}{\partial\beta''}[\beta''\delta(\beta''-\beta')]\}\, d\beta''$$

$$= -\frac{ih}{2\pi}\int \{(\beta-\beta'')\delta(\beta-\beta'')\delta'(\beta''-\beta') - \delta(\beta-\beta'')\delta(\beta''-\beta')\}\, d\beta''.$$

The first term vanishes since $x\delta x = 0$ and therefore

$$(\xi\eta - \eta\xi)(\beta\beta') = \frac{ih}{2\pi}\int \delta(\beta-\beta'')\delta(\beta''-\beta')\, d\beta''$$

$$= \frac{ih}{2\pi}\delta(\beta-\beta'); \text{ using (3) § 159.}$$

$$\therefore\ \xi\eta - \eta\xi = \frac{ih}{2\pi} 1.$$

163. *Transformation theory.*

We now transform ξ, η to the new matrix schemes of § 161. Using (4) § 161, we have

$$\xi(\beta\alpha) = \int \xi(\beta\beta')\, d\beta'(\beta'/\alpha)$$

$$= \int \beta\delta(\beta-\beta')\, d\beta'(\beta'/\alpha)$$

$$= \beta(\beta/\alpha), \text{ from (1) § 159.}$$

So
$$\eta\,(\beta\alpha) = \int \eta\,(\beta\beta')\,d\beta'\,(\beta'/\alpha)$$
$$= -\frac{ih}{2\pi}\int \delta'\,(\beta-\beta')\,d\beta'\,(\beta'/\alpha)$$
$$= -\frac{ih}{2\pi}\frac{\partial}{\partial\beta}\,(\beta/\alpha),\ \text{from (2) § 159.}$$

This suggests that if f is a function of ξ, η, rational and integral in η, $f(\xi,\eta)$ is a matrix, whose constituent
$$f(\xi,\eta)\,(\beta\alpha) = f\left(\beta, -\frac{ih}{2\pi}\frac{\partial}{\partial\beta}\right)(\beta/\alpha),$$
so that the elements of the matrix for f in the $(\beta\alpha)$ scheme are given by a certain operator operating upon (β/α).

It is sufficient to prove that if the theorem is true for each of two functions f_1 and f_2, it is true for their sum f_1+f_2 and their product $f_1 f_2$. The former is obvious. To see the latter we have

$$f_1(\xi,\eta)\,f_2(\xi,\eta)\,(\beta\alpha)$$
$$= \iint f_1(\xi,\eta)\,(\beta\alpha')\,d\alpha'\,(\alpha'/\beta')\,d\beta'\,f_2(\xi,\eta)\,(\beta'\alpha)$$
$$= \iint f_1\left(\beta, -\frac{ih}{2\pi}\frac{\partial}{\partial\beta}\right)(\beta/\alpha')\,d\alpha'\,(\alpha'/\beta')\,d\beta'$$
$$\times f_2\left(\beta', -\frac{ih}{2\pi}\frac{\partial}{\partial\beta'}\right)(\beta'/\alpha)$$
$$= f_1\left(\beta, -\frac{ih}{2\pi}\frac{\partial}{\partial\beta}\right) f_2\left(\beta, -\frac{ih}{2\pi}\frac{\partial}{\partial\beta}\right)(\beta/\alpha).$$

Thus has been proved
$$f(\xi,\eta)\,(\beta\alpha) = f\left(\beta, -\frac{ih}{2\pi}\frac{\partial}{\partial\beta}\right)(\beta/\alpha) \quad \text{.........(1).}$$

So it can be shown that
$$f(\xi,\eta)\,(\alpha\beta) = f\left(\beta, +\frac{ih}{2\pi}\frac{\partial}{\partial\beta}\right)(\alpha/\beta).$$

164. *Deduction of the Schrödinger wave equation.*

The result (1) of § 163 gives a powerful method of finding the matrix scheme which will make any function of the dynamical variables a diagonal matrix.

Suppose we are given a function H of the variables ξ, η and we want the matrix scheme (α) say, for which H will be a diagonal matrix $\equiv E$. Then

$$H(\alpha\alpha') = H(\alpha)\,\delta(\alpha - \alpha') = E_\alpha\,\delta(\alpha - \alpha'), \text{ suppose.}$$

Then

$$H\left(\beta, -\frac{ih}{2\pi}\frac{\partial}{\partial\beta}\right)(\beta/\alpha) = H(\xi, \eta)(\beta\alpha), \text{ from (1) § 163}$$

$$= \int (\beta/\alpha')\, d\alpha' H(\xi, \eta)(\alpha'\alpha), \text{ from (4) § 161}$$

$$= \int (\beta/\alpha')\, d\alpha' E_\alpha\,\delta(\alpha - \alpha')$$

$$= E_\alpha(\beta/\alpha), \text{ from (1) § 159.}$$

$$\therefore \quad \left\{H\left(\beta, -\frac{ih}{2\pi}\frac{\partial}{\partial\beta}\right) - E_\alpha\right\}(\beta/\alpha) = 0 \quad \text{......(1),}$$

which is a differential equation in β for (β/α). The different solutions are specified by different values of α.

If we write $\alpha = n$ and write (β/α), which

$$= S(\beta\alpha) = S(\beta n), \text{ as } S_n(\beta),$$

then
$$\left\{H\left(\beta, -\frac{ih}{2\pi}\frac{\partial}{\partial\beta}\right) - E_n\right\} S_n(\beta) = 0 \quad \text{.........(2).}$$

If we suppose ξ, η to be the ordinary q, p of the system at some specified time, and H to be the Hamiltonian, equation (1) is just *Schrödinger's wave equation* (§ 95). The eigenfunctions $S_n(\beta)$ or (β/α) are the elements of the transformation matrix S which enables one to find the matrix scheme in which the Hamiltonian is a diagonal matrix.

165. *Calculation of the matrix elements for any dynamical variable.*

For the dynamical variable $f(\xi, \eta)$,

$$f(\xi, \eta)(\alpha\alpha') = \int (\alpha/\beta)\, d\beta\, f(\xi, \eta)(\beta\beta')\, d\beta' (\beta'/\alpha'), \text{ from (3) §161}$$

$$= \int (\alpha/\beta)\, d\beta\, f(\xi, \eta)(\beta\alpha'), \text{ from (4) § 161}$$

$$= \int (\alpha/\beta)\, d\beta\, f\left(\beta, -\frac{ih}{2\pi}\frac{\partial}{\partial\beta}\right)(\beta/\alpha'), \text{ from (1) §163.}$$

Using the notation above that

$$(\beta/\alpha) = S(\beta\alpha) = S(\beta n) = S_n(\beta),$$

then

$$(\alpha/\beta) = S^{-1}(\alpha\beta) = S^{-1}(n\beta) = \tilde{S}^{-1}(\beta n) = S^*(\beta n) = S_n^*(\beta),$$

since $S^{-1} = \tilde{S}^*$, on account of $S\tilde{S}^* = 1$.

$$\therefore\ f(\xi, \eta)(nm) = \int S_n^*(\beta) f\left(\beta, -\frac{ih}{2\pi}\frac{\partial}{\partial\beta}\right) S_m(\beta)\, d\beta \quad \ldots(1).$$

The S_n's are found from the wave equation (1) § 159 and the matrix elements for $f(\xi, \eta)$ follow from (1) by a process of integration.

166. *Summary.*

The solution of a problem in quantum mechanics is the solution of the eigenwert problem $SH(p, q) S^{-1} = E$; $S\tilde{S}^* = 1$, where $H(p, q)$ is the Hamiltonian, E the diagonal matrix for the energy, and S is the transformation matrix sought for.

A new system of canonical variables ξ, η is introduced such that the ξ's are all diagonal matrices. In S, the constituents are $S(n\beta)$ or $S_n(\beta)$.

The equation $SHS^{-1} = E$ proves to be equivalent to

$$\left\{H\left(\beta, -\frac{ih}{2\pi}\frac{\partial}{\partial\beta}\right) - E_n\right\} S_n(\beta) = 0,$$

whence $S_n(\beta)$ can be found.

Also

$$\int S_n(\beta) S_m^*(\beta) d\beta = \delta_{nm} \begin{array}{l} = 1 \text{ for } n = m \\ = 0 \text{ for } n \neq m. \end{array}$$

The matrix elements for a function $f(\xi, \eta)$ are then given by

$$f(\xi, \eta)(nm) = \int S_n^*(\beta) f\left(\beta, -\frac{ih}{2\pi}\frac{\partial}{\partial\beta}\right) S_m(\beta) d\beta.$$

Jordan[1] has independently worked out an operator theory which leads to the same results.

The Dirac-Jordan theory is a comprehensive one which includes the older Heisenberg-Dirac matrix theory, the operator theories of Lanczos[2] and of Born and Wiener[3], and the wave equation of Schrödinger.

167. *Generalised form of the Schrödinger equation (Dirac*[4]*).*

If the Hamiltonian H is given as a function of ξ, η, then since

$$\left.\begin{array}{ll} & \xi(\beta\beta') = \beta\delta(\beta - \beta') \\ \text{and} & \eta(\beta\beta') = -\dfrac{ih}{2\pi}\delta'(\beta - \beta'), \ (\S\ 162) \end{array}\right\} \quad \ldots\ldots(1),$$

the matrix $H(\beta\beta')$ is known.

Hence using (4) § 161,

$$\int H(\beta\beta') d\beta' (\beta'/\alpha) = \int (\beta/\alpha') d\alpha' H(\alpha'\alpha).$$

If H is to be a diagonal matrix in the α's, then

$$H(\alpha\alpha') = H(\alpha)\,\delta(\alpha - \alpha').$$

$$\therefore \quad \int H(\beta\beta') d\beta' (\beta'/\alpha) = \int (\beta/\alpha') d\alpha' H(\alpha')\,\delta(\alpha' - \alpha)$$

$$= (\beta/\alpha) H(\alpha), \text{ using (1) § 159.}$$

[1] P. JORDAN, Zs. f. Phys. **40**, p. 809, **1927**; **44**, p. 1, **1927**, and earlier papers in the Zs. f. Phys.

[2] K. LANCZOS, Zs. f. Phys. **35**, p. 812, **1926**.

[3] M. BORN and N. WIENER, Zs. f. Phys. **36**, p. 174, **1926**.

[4] P. A. M. DIRAC, Proc. Roy. Soc. A. **114**, p. 243, **1927**.

We thus have

$$\int H(\beta\beta')(\beta'/\alpha)\,d\beta' = (\beta/\alpha)\,H(\alpha) \quad \text{.........(2)},$$

which is an *integral equation* for (β/α) of which the characteristic values $H(\alpha)$ are the energy levels.

Schrödinger's wave equation is a special case of (2) where H is an *algebraic function* of ξ, η. It then becomes an ordinary differential equation on account of the relations (1), and takes the form of (1) § 160.

It is possible to have a system specified by a Hamiltonian H which cannot be expressed as an algebraic function of a set of canonical variables, but which yet can be represented by a matrix whose terms are $H(\beta\beta')$. Equation (2) can then be used to find the energy levels and the eigenfunctions. The Hamiltonian for the interaction of a light quantum and an atomic system is of this more general type.

168. *Physical interpretation of q-numbers.*

Any q-number can be expressed as a diagonal matrix if the matrix system is suitably chosen and the diagonal terms are its 'characteristic values'; these represent ordinary numerical values which can be compared with experiment.

Suppose x, y are the action and angle variables, for example, of a system of one freedom and that the rows and columns of the matrices are designated by the characteristic values of x, namely $\alpha, \alpha', \alpha'', \ldots$, where x is the diagonal matrix $\begin{pmatrix} \alpha, & 0 \ldots \\ 0, & \alpha' \ldots \\ \ldots & \ldots \end{pmatrix}$.

Let ξ, η be canonical variables, functions of x, y, and let the matrices be transformed so that ξ is the diagonal matrix $\begin{pmatrix} \beta, & 0 \ldots \\ 0, & \beta' \ldots \\ \ldots & \ldots \end{pmatrix}$ in the new scheme.

Given $\xi = f(x, y)$, we wish to find what ordinary numerical relations we can between ξ, x, y. We can give x one of its characteristic values α, say, but we cannot give y one of its characteristic values because y cannot be a diagonal matrix at the same time as x on account of the quantum condition $yx - xy = \frac{ih}{2\pi} 1$. But it is possible to find the *range* of values of y for which ξ has the characteristic value β, say, when x has the given characteristic value α.

Dirac[1] has shown that this is the fraction $(\alpha/\beta)(\beta/\alpha)$ of the whole y range.

Thus for an assembly of atoms all having the same numerical value of α (i.e. all in the same stationary state) and uniformly distributed as regards y, the number of atoms for which ξ has the numerical value β is $(\alpha/\beta)(\beta/\alpha)$ of the whole number.

Thus for a set of atoms for which x is α we can deduce only a probability function $(\alpha/\beta)(\beta/\alpha)$ for the number of atoms for which ξ will be β.

If the numbers β are continuous and not discrete, the probability for the number of atoms for which ξ lies between β and $\beta + d\beta$ is $(\alpha/\beta)\, d\beta\, (\beta/\alpha)$.

In the notation of § 165, this probability is

$$S_n^* (\beta)\, S_n (\beta)\, d\beta \quad \text{or} \quad | S_n (\beta) |^2 d\beta;$$

this is the probability that in the stationary state n the variable ξ lies between β and $\beta + d\beta$.

This form was used by Born in his papers on collision phenomena, referred to earlier (p. 7).

This replacement of a state by a probability of the system having that state is a characteristic feature of quantum dynamics, which has recently been emphasised by Heisenberg[2] on general physical grounds (chap. XXXII).

[1] P. A. M. Dirac, Proc. Roy. Soc. A. **113**, p. 621 (§ 6), **1927.**
[2] W. Heisenberg, Zs. f. Phys. **43**, p. 172, **1927.**

169. *Successive matrix transformations.*

If we apply successively two canonical transformations with the matrices S, T, so that

$$G = SgS^{-1},$$
$$G' = TGT^{-1},$$

then $$G' = TSgS^{-1}T^{-1} = TSg\,(TS)^{-1}.$$

Thus the result is the same as the single transformation with the matrix TS.

In the new notation two successive transformations from scheme α to scheme β and from scheme β to scheme γ are equivalent to a direct transformation from α to γ, provided

$$\left.\begin{aligned}(\gamma/\alpha) &= \int (\gamma/\beta)\, d\beta\, (\beta/\alpha)\\ \text{and}\qquad (\alpha/\gamma) &= \int (\alpha/\beta)\, d\beta\, (\beta/\gamma)\end{aligned}\right\}\ldots\ldots\ldots\ldots\ldots(1).$$

170. *Application to collision problems.*

Consider the case of collisions between an electron and an atomic system. In Born's[1] treatment of the problem, a solution of the wave equation is found consisting of incident plane waves representing the oncoming electron. These waves are scattered by the atomic system. It is then assumed that the square of the amplitude of the wave scattered in any direction determines the probability of the electron being scattered in that direction, with an energy given by the frequency of the wave.

By Dirac's method, the transformation function (p_t/p_0) is found which connects the momentum at time t (expressed as a diagonal matrix with components p_t) with the initial momentum (expressed as a diagonal matrix with components p_0). [The p_t, p_0 correspond to the β, α of the above theory.] There is then a probability

$$(p_0/p_t)\, dp_t\, (p_t/p_0) \text{ or } |\,(p_t/p_0)\,|^2\, dp_t$$

[1] M. Born, Zs. f. Phys. **37**, p. 863; **38**, p. 803, **1926**. See also P. Dirac, Zs. f. Phys. **44**, p. 585, **1927**.

that the electron will be scattered with a momentum lying in the range dp_t.

If x_t is a coordinate of the electron at time t, then by (1), § 165,

$$(x_t/p_0) = \int (x_t/p_t)\, dp_t\, (p_t/p_0) \quad \text{...........(1)}.$$

Here the transformation function x_t/p_0 is the solution of the wave equation appropriate to the case of an incident electron whose momentum is p_0 and is thus the wave function of Born's theory.

The function x_t/p_t represents emerging waves corresponding to electrons with momentum p_t.

Thus equation (1) gives the resolution of the emerging waves in the eigenfunction x_t/p_t into their different components, the amplitudes of the various components being $|(p_t/p_0)|$. This agrees with Born's theory quoted above.

It is thus seen that in general the dynamical variables can be represented equally well by matrices whose rows and columns refer to the initial values of the action variables (p_0) or to the final values (p_t). The transformation function from the one set of matrices to the other is the associated probability.

CHAPTER XXXII

THE ESSENTIAL INDEFINITENESS OF QUANTUM MECHANICS; HEISENBERG'S PHILOSOPHY AND BOHR'S DE BROGLIE WAVE THEORY OF THE PASSAGE FROM MICRO- TO MACRO-MECHANICS; THE SPREADING FACTOR; BOHR'S SUMMARY OF THE PRESENT STATE OF THE QUANTUM THEORY

171. *The essential indefiniteness of quantum mechanics*[1] (*Heisenberg*).

The origin of quantum mechanics was an attempt to break away from the usual kinematical and mechanical concepts and in their place to substitute relations between concrete numbers given by experiment.

(*a*) *The position of an electron.*

The position of an electron relative to a given system of axes must be measurable by experiment if the term is to have a meaning in quantum mechanics. We illuminate the electron and observe it through a microscope; the accuracy with which the position can be observed is given by the wave-length of the light used. If we use a γ *ray microscope*, we should get the most exact result physically possible. But these γ rays produce a Compton effect on the electron, so that at the instant when the position is determined the momentum of the electron is suddenly changed. This change of momentum is all the greater the shorter the wave-length of the light used, i.e. the more accurate the determination of position is.

[1] W. Heisenberg, Zs. f. Phys. **43**, p. 172, **1927**.

Thus at the moment when the position of the electron is known, its momentum can only be known to an order equal to this discontinuous change. Hence the more exactly a coordinate q is determined, the less exactly can its momentum p be found; and conversely. This is a direct intuitive illustration of the meaning of the quantum condition

$$qp - pq = \frac{ih}{2\pi} 1;$$

for if q_1 is the exactness with which the value of q is known (so that q_1 is the mean error of q, here the wave-length of the light used) and p_1 is the mean error for p, then

$$p_1 q_1 \sim h \quad \text{.......................(1).}$$

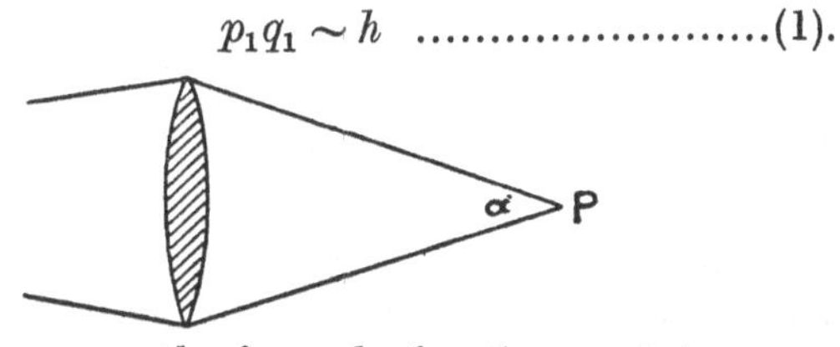

To see this we use the formula for the resolving power of a microscope[1]; the distance between two distinguishable points at P must $> \lambda/\alpha$, where λ is the wave-length of the γ rays, or

$$\Delta q \sim \frac{\lambda}{\alpha}.$$

The momentum p of a light quantum $= \frac{h\nu}{c} = \frac{h}{\lambda}$, and since the direction of p may lie anywhere in a range of angle α, the possible error of p is $p\alpha$, or $\Delta p \sim p\alpha$,

$$\therefore \ \Delta p \sim p\alpha \sim \frac{h\alpha}{\lambda},$$

$$\therefore \ \Delta p . \Delta q \sim \frac{h\alpha}{\lambda} . \frac{\lambda}{\alpha} \sim h,$$

or

$$p_1 q_1 \sim h.$$

[1] P. DRUDE, 'The Theory of Optics,' English translation, Longmans, 1908.

This result can also be found by the use of de Broglie waves.

If Δq is the range within which q can be found by experiment, Δq is of the order of the wave-length l of the de Broglie group associated with the electron. If λ is the length of a wave, then $\frac{1}{l} = \frac{\Delta\lambda}{\lambda^2}$ from the usual group relation, so that

$$\Delta\lambda \sim \frac{\lambda^2}{\Delta q} \quad \text{..........................(2).}$$

But the momentum $p = \frac{h\nu}{c} = \frac{h}{\lambda}$, so that

$$|\Delta p| = \frac{h\,|\Delta\lambda|}{\lambda^2} \quad \text{..................(3).}$$

Therefore from (2) and (3), $(\Delta p)\,(\Delta q) \sim h$, which is the result (1).

(*b*) *The path of an electron.*

The familiar expression 'the $1s$ path of an electron' has from this point of view no meaning. For in order to measure this path (which in the Bohr theory for hydrogen is of order 10^{-9} cm.)[1] we must illuminate it with light whose wave-length is far shorter than 10^{-9} cm. One quantum of such light would suffice to drive the electron from its orbit to some other orbit, so that on this account only one point of the '$1s$ path' can be observed and the word path has no rational meaning.

But on the other hand the 'position' measurement of a *system* of atoms in the $1s$ state can be measured, as is known by the Stern-Gerlach experiments (§ 17). Thus for a given state $1s$ of the atoms these experiments determine a 'probability function' for the position of the electron, which corresponds to the 'mean value' of the position in a classical path. According to Born[2] this

[1] Q.T.A. § 18.
[2] M. Born, Zs. f. Phys. **38**, p. 803, **1926**.

function is equal to $\psi_{1s}(q)\,\overline{\psi}_{1s}(q)$, where $\psi_{1s}(q)$ is the Schrödinger wave function corresponding to the state 1*s* and $\overline{\psi}$ is the conjugate of ψ.

Dirac and Jordan (§ 164) would say 'the probability is $|\beta/\alpha|^2$, where β/α is that term of the transformation matrix $S(\beta\alpha)$ which transforms the matrices where E is a diagonal matrix in the α's to matrices where q is a diagonal matrix in the β's.' Here α is the energy of the state 1*s*, or $\alpha = E_{1s}$.

This characteristic indefiniteness or statistical trait of quantum mechanics is in strong contrast with the concepts of the classical mechanics. In the latter the position of a particle at *each* point of its orbit can be found; in the former it cannot, as the determination of *one* point distorts the atom and only a probability function for the position of an electron in a given stationary state can be found.

Then again the quantum condition

$$Et - tE = \frac{ih}{2\pi}\,1 \ (\S\ 104),$$

suggests $E_1 t_1 \sim h$, where E_1, t_1 are mean errors in E, t. Thus the more exactly the energy of a stationary state is given, the less exact is t, so that the phases are indeterminate. Also one cannot speak of the energy at a given time, as precision in t means vagueness in E; such indefiniteness must always arise in the experimental determination of any pair of canonical conjugate variables.

Heisenberg proceeds to develop this idea $p_1 q_1 \sim h$ with a wealth of illustration derived from very diverse experimental data. He then goes on to show that $p_1 q_1 = \dfrac{h}{2\pi}$ by the use of the Dirac-Jordan theory.

Bohr has, however, lately obtained this and other results of Heisenberg by the use of a calculus (§§ 172–5) which he has not yet had time to publish, and which is founded upon the de Broglie wave theory.

172[1]. *Fourier theorems for the de Broglie waves.*

An ordinary Fourier series for $f(q)$ is $\Sigma A_\lambda e^{\frac{2\pi i q}{\lambda}}$, where λ is the wave-length in q; or as a Fourier integral

$$f(q) = \int F(\lambda)\, e^{\frac{2\pi i q}{\lambda}}\, d\lambda.$$

For de Broglie waves, the momentum $p = \frac{h\nu}{c} = \frac{h}{\lambda}$, so that $\lambda = h/p$.

Hence
$$\psi(q) = \int_{-\infty}^{\infty} g(p)\, e^{\frac{2\pi i p q}{h}}\, dp \quad \ldots\ldots\ldots\ldots(1).$$

Again, Fourier's integral translates into

$$F(p) = \frac{1}{h}\int dq \int dp' F(p') \exp \frac{2\pi i (p-p') q}{h} \quad \ldots(2).$$

We can now invert (1) by writing it as

$$\psi(q) = \int_{-\infty}^{\infty} g(p')\, e^{\frac{2\pi i p' q}{h}}\, dq',$$

multiplying both sides by $e^{-\frac{2\pi i p q}{h}}$, and integrating with respect to q from $-\infty$ to ∞.

Then

$$\int dq \psi(q) \exp\left(-\frac{2\pi i p q}{h}\right) = \int dq \int_{-\infty}^{\infty} dp' g(p') \exp \frac{2\pi i (p'-p) q}{h}$$
$$= hg(p), \text{ using (2).}$$

Hence (1) inverts into

$$g(p) = \frac{1}{h}\int_{-\infty}^{\infty} \psi(q)\, e^{-\frac{2\pi i p q}{h}}\, dp \quad \ldots\ldots\ldots\ldots(3).$$

[The limits $-\infty$ to ∞ will be omitted from now.]

173. *Bohr's fundamental probability theorem.*

Let $\psi_n(q)$ be the Schrödinger wave function for an electron $(-\epsilon)$ in a stationary state defined by n, where q is a position coordinate. We will write this just $\psi(q)$, as n

[1] The theory of §§ **172–5** was communicated to the author when at Copenhagen in September, **1927**.

does not enter into the calculations to follow. Suppose ψ normalised, so that $\int \psi(q)\,\bar{\psi}(q)\,dq = 1$, where $\bar{\psi}$ is the conjugate of ψ; or what is the same

$$\int |\,\psi(q)\,|^2\,dq = 1 \quad \text{..................(1).}$$

The electron $(-\epsilon)$ is equivalent to a Schrödinger electric density $\psi(q)\,\bar{\psi}(q)$ throughout space (§ 103), and since

$$-\epsilon = -\int \epsilon\psi\bar{\psi}\,dq, \text{ from (1),}$$

we may take $-\epsilon\psi\bar{\psi}$ as the space charge associated with q; or otherwise expressed, $\psi\bar{\psi}$ is the probability of the electron being in the place q.

This agrees with the Dirac result of § 168; it is found in a more intuitive way.

Bohr now expands $\psi(q)$ as a de Broglie Fourier integral, so that

$$\psi(q) = \int g(p)\,e^{\frac{2\pi ipq}{h}}\,dp \quad \text{..............(2),}$$

and shows that if $|\,\psi(q)\,|^2$ is the probability for the electron to have the place q, then $h\,|\,g(p)\,|^2$ is the probability for it to have the momentum p.

For

$$1 = \int \psi(q)\,\bar{\psi}(q)\,dq$$

$$= \int dq \int g(p)\,e^{\frac{2\pi ipq}{h}}\,dp \int \bar{g}(p')\,e^{-\frac{2\pi ip'q}{h}}\,dp',$$

substituting for ψ from (2),

$$= \int dp\,g(p) \int dq \int dp'\,\bar{g}(p')\,e^{\frac{2\pi i}{h}(p-p')q}$$

$$= h\int dp\,g(p)\,\bar{g}(p), \text{ using (2), § 172.}$$

Thus $hg(p)\,\bar{g}(p) \equiv h\,|\,g(p)\,|^2$ is the probability for the p.

The relations (1), (3), § 172, enable either of $\psi(q)$, $g(p)$ to be found when the other is given.

174. *The theorem* $p_1 q_1 = \frac{h}{2\pi}$.

If p is known with an exactness p_1 (p_1 is the 'mean error' of p) and q is known with an exactness q_1, then $p_1 q_1 = \frac{h}{2\pi}$, where p, q are two canonically conjugate variables.

Suppose the probability $\psi(q)\bar{\psi}(q)$ of q to be given by a Gauss error curve of effective range q_1, so that

$$\psi(q)\bar{\psi}(q) = e^{-\frac{q^2}{q_1^2}}.$$

Then
$$\psi(q) = e^{-\frac{q^2}{2q_1^2}+iQ},$$
where Q is any real function of q.

Choose Q to be linear in q and equal to $\frac{2\pi}{h}\lambda q$, where λ is any constant.

Then
$$\psi(q) = \exp\left(-\frac{q^2}{2q_1^2} + \frac{2\pi i}{h}\lambda q\right)$$

and
$$g(p) = \frac{1}{h}\int \psi(q) \exp\left(-\frac{2\pi i p q}{h}\right) dp, \text{ § 172 (3)},$$
$$= \frac{1}{h}\int \exp\left(-\frac{q^2}{2q_1^2} + \frac{2\pi i}{h}\lambda q - \frac{2\pi i p q}{h}\right) dp.$$

Using the result

$$\int_{-\infty}^{\infty} \exp(-ax^2 - 2bx)\, dx = \sqrt{\frac{\pi}{a}} \exp\left(\frac{b^2}{a}\right) \ldots (1),$$

$$g(p) = \frac{1}{h}\sqrt{\pi}\sqrt{2}\, q_1 \exp\left\{-\frac{\pi^2}{h^2}(p-\lambda)^2\, 2q_1^2\right\},$$

$$\therefore\ g(p) = \frac{\sqrt{2\pi}}{h} q_1 \exp\left\{-\frac{(p-\lambda)^2}{2p_1^2}\right\} \ldots\ldots\ldots (2),$$

where
$$\frac{1}{2p_1^2} = \frac{2\pi^2 q_1^2}{h^2},$$

or
$$p_1 q_1 = \frac{h}{2\pi},$$

$$\therefore\ hg(p)\bar{g}(p) = \frac{2\pi q_1^2}{h} \exp\left\{-\frac{(p-\lambda)^2}{p_1^2}\right\}.$$

Thus the p probability $hg(p)\,\bar{g}(p)$ is represented by a Gauss error curve of range p_1 about $p = \lambda$.

If then we suppose that when q is in a range q_1 about $q = 0$, that p is in a range p_1 about $p = p_0$, then the above show that

$$\psi(q) = \exp\left(-\frac{q^2}{2q_1^{\,2}} + \frac{2\pi i}{h} p_0 q\right)$$

and
$$g(p) = \frac{\sqrt{2\pi}}{h} q_1 \exp\left\{-\frac{(p-p_0)^2}{2p_1^{\,2}}\right\},$$

with
$$p_1 q_1 = \frac{h}{2\pi}.$$

175. *The passage from micro- to macro-mechanics (Heisenberg-Bohr).*

Let $\psi(q, t_0)$ be the wave function, so that

$$\psi(q, t_0)\,\bar{\psi}(q, t_0)$$

is the probability function for the position q at time t_0 of the electron considered. This 'probability packet' corresponds to a Schrödinger 'wave packet' at time t_0.

Heisenberg considers the case of uniform rectilinear motion (with no external field) for which the Hamiltonian $H = \frac{p^2}{2\mu}$, where μ is the mass.

The wave equation $(H - E)\,\psi = 0$ is here

$$\left\{\frac{1}{2\mu}\left(-\frac{ih}{2\pi}\frac{\partial}{\partial q}\right)^2 - \frac{ih}{2\pi}\frac{\partial}{\partial t}\right\}\psi = 0 \text{ (§ 95)},$$

or
$$-\frac{\partial^2\psi}{\partial q^2}\frac{ih}{4\pi\mu} + \frac{\partial\psi}{\partial t} = 0.$$

Write
$$\psi = ue^{-\frac{2\pi i E t}{h}},$$

then
$$-\frac{d^2 u}{dq^2}\frac{ih}{4\pi\mu} - \frac{2\pi i E}{h} u = 0.$$

Putting in $E = \frac{p^2}{2\mu}$, this becomes

$$\frac{d^2u}{dq^2} + \frac{4\pi^2 p^2}{h^2} u = 0,$$

$$\therefore \quad u = C \exp \frac{2\pi}{h} pqi,$$

$$\therefore \quad \psi = C \exp \frac{2\pi i}{h} (pq - Et),$$

or
$$\psi = C \exp \frac{2\pi i}{h} \left(pq - \frac{p^2 t}{2\mu}\right).$$

Thus if ψ were $\exp \frac{2\pi i}{h} pq$ at time $t = 0$, it would become

$$\exp \frac{2\pi i}{h} \left(pq - \frac{p^2 t}{2\mu}\right) \text{ at time } t = t \quad \text{.........(1).}$$

Using the results of § 174, and supposing that q is in a range q_1 about $q = 0$ at time $t = 0$,

$$\psi(q, t_0) = \exp\left(-\frac{q^2}{2q_1^2} + \frac{2\pi i}{h} p_0 q\right).$$

Also
$$g(p, t_0) = \frac{1}{h} \int \psi(q, t_0) \exp\left(-\frac{2\pi i p q}{h}\right) dq.$$

Therefore using (1),

$$g(p,t) = \frac{1}{h} \int \psi(q, t_0) \exp\left\{-\frac{2\pi i}{h}\left(pq - \frac{p^2 t}{2\mu}\right)\right\} dq$$

$$= \frac{1}{h} \int \exp\left\{-\frac{q^2}{2q_1^2} + \frac{2\pi i}{h} p_0 q - \frac{2\pi i}{h}\left(pq - \frac{p^2 t}{2\mu}\right)\right\} dq$$

$$= \frac{\sqrt{2\pi}}{h} q_1 \exp\left\{-\frac{\pi^2 (p - p_0)^2}{h^2} 2q_1^2 + \frac{\pi i p^2 t}{h\mu}\right\},$$

using (1), § 174.

$$\therefore \quad g(p, t) = \sqrt{2\pi} \frac{q_1}{h} \exp - \frac{(p - p_0)^2}{2p_1^2} + \frac{\pi i p^2 t}{h\mu},$$

since
$$p_1 q_1 = \frac{h}{2\pi} \quad \text{........................(2).}$$

But $$\psi(q,t) = \int g(p,t) \exp \frac{2\pi ipq}{h}\, dp,$$

$$\therefore\ \psi(q,t) = \sqrt{2\pi}\,\frac{q_1}{h} \int \exp\left\{-\frac{(p-p_0)^2}{2p_1^{\,2}} + \frac{\pi ip^2t}{h\mu} + \frac{2\pi ipq}{h}\right\} dp$$

$$= \sqrt{2\pi}\,\frac{q_1}{h}\,\frac{\sqrt{\pi}}{\sqrt{\dfrac{1}{2p_1^{\,2}} - \dfrac{\pi it}{h\mu}}} \exp\left[\frac{\left(\dfrac{\pi iq}{h} + \dfrac{p_0}{2p_1^{\,2}}\right)^2}{\dfrac{1}{2p_1^{\,2}} - \dfrac{\pi it}{h\mu}} - \frac{p_0^{\,2}}{2p_1^{\,2}}\right].$$

Writing $\beta = \dfrac{2\pi tp_1^{\,2}}{h\mu}$ and using $p_1q_1 = \dfrac{h}{2\pi}$,

$$\psi(q,t) = \frac{1}{\sqrt{1-i\beta}} \exp\left[\left\{-q^2 + \frac{ih}{\pi}\left(\frac{tp_0^{\,2}}{2p_1^{\,2}\mu} - \frac{qp_0}{p_1^{\,2}}\right)\right\} \Big/ 2q_1^{\,2}(1-i\beta)\right]$$

$$= \frac{1}{\sqrt{1-i\beta}} \exp\left\{-\frac{q^2+i\lambda}{1-i\beta}\cdot\frac{1}{2q_1^{\,2}}\right\},$$

where $$\lambda = \frac{h}{\pi p_1^{\,2}}\left(\frac{tp_0^{\,2}}{2\mu} - qp_0\right).$$

Therefore the q-probability at time t, which is

$$\bar{\psi}(q,t)\,\psi(q,t), = \frac{1}{\sqrt{1+\beta^2}} \exp\left\{-\frac{1}{2q_1^{\,2}}\left(-\frac{q^2+i\lambda}{1-i\beta} + \frac{-q^2-i\lambda}{1+i\beta}\right)\right\}$$

$$= \frac{1}{\sqrt{1+\beta^2}} \exp\left\{-\frac{(q^2+\lambda\beta)}{2q_1^{\,2}(1+\beta^2)}\right\}.$$

But $$q^2 + \lambda\beta = q^2 + \frac{2t}{\mu}\left(\frac{tp_0^{\,2}}{2\mu} - qp_0\right)$$

$$= \left(q - \frac{p_0t}{\mu}\right)^2.$$

Therefore the q-probability at time t

$$= \frac{1}{\sqrt{1+\beta^2}} \exp\left\{-\frac{\left(q - \dfrac{p_0t}{\mu}\right)^2}{2q_1^{\,2}(1+\beta^2)}\right\}.$$

Thus q has a range $q_1\sqrt{1+\beta^2}$ about the place

$$q = \frac{p_0 t}{\mu} = vt,$$

where v is the velocity p_0/μ.

Thus the size of the 'wave packet' or 'probability packet' is increased in the ratio $\sqrt{1+\beta^2}$, where

$$\beta = \frac{2\pi p_1{}^2 t}{h\mu}.$$

This 'spreading factor' $\sqrt{1+\beta^2}$ increases indefinitely with the time[1].

Thus usually a wave packet will diffuse itself indefinitely and not behave as suggested by Schrödinger in § 98.

The reason for this spreading can be seen on general grounds. For if there is a place-error q_1 at time $t=0$, there is also a place-error $\frac{p_1 t}{\mu}$ arising from doubt about p at time $t=0$. Thus the place-error $q_1(t)$ at time t, by the law of squares, is given by

$$\{q_1(t)\}^2 = q_1{}^2 + \left(\frac{p_1 t}{\mu}\right)^2.$$

Since
$$p_1 q_1 = \frac{h}{2\pi},$$

$$\begin{aligned}\{q_1(t)\}^2 &= q_1{}^2\left\{1 + \frac{h^2t^2}{4\pi^2\mu^2 q_1{}^4}\right\} \\ &= q_1{}^2\left\{1 + \frac{4\pi^2 p_1{}^4 t^2}{h^2\mu^2}\right\} \\ &= q_1{}^2(1+\beta^2).\end{aligned}$$

176. *Bohr's[2] summary of the present state of the quantum theory.*

The recent developments of atomic theory due to the de Broglie-Schrödinger outlook have led to a renewal of

[1] See also E. H. KENNARD, Zs. f. Phys. **44**, p. 326, **1927**.
[2] N. BOHR, Nature (to appear shortly).

the discussion[1] of the need for the postulates of the quantum theory hitherto used to coordinate experimental results.

Schrödinger discovered that the stationary states of an atom can be represented by the proper vibrations of a certain wave problem which replaces the particle problem of classical dynamics. Tracing the consequences of superposing such vibrations, he has been able to reproduce fundamental features of atomic processes representing transitions between stationary states in the language of the quantum postulates. In view of these results, he expresses the hope that a development of the wave theory would make it possible to *completely avoid* the irrational element involved in these postulates, and gradually build up a description of atomic phenomena on the lines of the classical physical theories. In reply to this Bohr asserts that in any process open to direct physical observation there is an essential discontinuity, or rather *individuality*, completely foreign to classical ideas and symbolised by Planck's constant h. This means an abandonment of causal space-time coordination of atomic phenomena and reduces the description of these phenomena to a statistical one.

It is possible to secure an asymptotic convergence to the classical theory whenever in statistical applications the discontinuities may be neglected. This is the *correspondence principle* which stands as the endeavour, in spite of the fundamental difference, to regard the quantum theory as a natural generalisation of the classical theory.

The assumption of a continuous character of collision and radiation processes in connection with the possibility of simultaneous excitation of different proper vibrations contemplated by Schrödinger would appear definitely to contradict the experiments on the excitation of spectral lines.

[1] E. Schrödinger, Ann. der Phys. **83**, p. 956, **1927**.

According to the quantum theory any observation depends upon individual processes which involve an essential interaction; the classical space-time coordination rests upon tools of measurement which do not affect the phenomena observed. The quantum theory replaces the causal space-time coordination of classical physics by two apparently contradictory but actually complementary ideas of individuality and superposition[1].

[1] These questions have lately been fully discussed at the Solvay Congress in Brussels (October **1927**).

INDEX OF AUTHORS

The numbers refer to the pages

CAMBRIDGE: PRINTED BY W. LEWIS, M.A., AT THE UNIVERSITY PRESS

Printed in the United States
By Bookmasters